'Just' a Fisherman's Wife

‘Just’ a Fisherman’s Wife: A Post Structural Feminist Exposé of Australian Commercial Fishing Women’s Contributions and Knowledge, ‘Sustainability’ and ‘Crisis’

By

Jane Dowling

CAMBRIDGE
SCHOLARS
PUBLISHING

'Just' a Fisherman's Wife:
A Post Structural Feminist Exposé of Australian Commercial Fishing Women's Contributions and Knowledge, 'Sustainability' and 'Crisis',
by Jane Dowling

This book first published 2011

Cambridge Scholars Publishing

12 Back Chapman Street, Newcastle upon Tyne, NE6 2XX, UK

British Library Cataloguing in Publication Data
A catalogue record for this book is available from the British Library

ISBN (10): 1-4438-2943-9, ISBN (13): 978-1-4438-2943-4

TABLE OF CONTENTS

LIST OF FIGURES

LIST OF TABLES

Acknowledgements

I would firstly like to acknowledge the amazing women who were involved in this research. Without your invaluable participation, enthusiasm and support, this research would not have been possible. Thank you to each of you, and I hope that this research proves to be useful for women in fishing.

I would also like to thank: the Women's Industry Network Seafood Community (WINSC); the Bureau of Rural Sciences (BRS); Charles Sturt University (CSU); the Australian Research Council (ARC); Professor Margaret Alston; Narrawallee Creek Studio, and, of course, my very special family, partner and amazing children.

ABBREVIATIONS

ABARE	Australian Bureau of Agricultural and Resource Economics
ABS	Australian Bureau of Statistics
AFFA	Agriculture, Fisheries and Forestry Australia
AFMA	Australian Fisheries Management Authority
APAI	Australian Postgraduate Award (Industry)
ARLP	Australian Rural Leadership Program
ARC	Australian Research Council
ASIC	Australian Seafood Industry Council
AWiA	Australian Women in Agriculture
BAS	Business Activity Statement
BRS	Bureau of Rural Sciences
CSU	Charles Sturt University
DPIE	Department of Primary Industries and Energy
DPIWE	Department of Primary Industries, Water and Environment (Tas)
ESD	Ecologically Sustainable Development
FAO	Food and Agriculture Organisation
FRDC	Fisheries Research and Development Corporation
FSRS	Fishermen and Scientists Research Society
GDP	Gross Domestic Product
ITQ	Individual Transferable Quota
MAC	Management Advisory Committee
MSC	Marine Stewardship Council
NAC	National Aquaculture Council
NCP	National Competition Policy
CALD	Culturally and Linguistically Diverse
NFFC	National Farmers Federation Council
NRE	National Resources and Environment
NSW	New South Wales
RIRDC	Rural Industries Research and Development Corporation
RLC	Western Rock Lobster Council
RLIAC	Rock Lobster Industry Advisory Council
RWAC	Regional Women's Advisory Council
RWS	Rural Women's Secretariat
SCARM	Standing Committee on Agriculture and Resource Management
TAC	Total Allowable Catch
TAFE	Technical and Further Education
TAS	Tasmania
TFIC	Tasmanian Fishing Industry Council
UN	United Nations

WA	Western Australian
WAFIC	Western Australian Fishing Industry Council
WCED	World Commission on Environment and Development
WINSC	Women's Industry Network Seafood Community
WIRIS	Women in Rural Industries and Industry Leadership Section

INTRODUCTION

Both the fishing and agricultural industries are traditionally viewed as a masculine domain, one in which women are mostly invisible. Yet, there is a lack of research on women in commercial fishing industry in Australia. This does not necessarily reflect a lack of women involved in the industry. Rather, it identifies the need to investigate the possibility that the Australian commercial fishing industry is locally, structurally, politically and systemically gendered.

This book provides a snapshot of women in family fishing businesses that operate licenses and leases in both the aquaculture and wild catch sectors. It applies a feminist post structural framework to explore the lack of visibility of these women in fishing; their contributions; the barriers to, and opportunities for, their greater participation; and the implications of their industry positioning in the quest for more sustainable fisheries in the current industry "crisis". As these issues frame the boundaries of the research, this book does not provide extensive technical information about harvesting practices or a comprehensive history of fisheries management. Instead, these issues are explained where relevant.

This book is based on doctoral research undertaken between 2003 and 2006 under a Australian Postgraduate Award (Industry) (APAI) scholarship and in partnership between the Centre for Rural Social Research-Charles Sturt University (CSU) and the Bureau of Rural Sciences (BRS). Given the time lapse between conducting the research and the publication of this book, the context in which this research was undertaken has evolved, particularly in relation to fisheries management policies and regulations, and other environmental, economic and political factors. There has also been a change of Commonwealth government. While any such changes potentially may be associated with shifting of gender relations, identities and positioning, recent publications suggest that there has been little change in terms of women's marginalised position in both fishing (e.g. Weeratunge and Snyder 2009) and agriculture (e.g. Sheridan and Haslam-McKenzie 2009). As such, it can be said with some certainty that the pervasiveness of the gendered nature of the Australian fishing industry is such that industry women continue to remain marginalised.

Furthermore, it has been acknowledged that while progress has been made in terms of implementing Ecologically Sustainable Development (ESD) in Australian fisheries, the focus has been on ecological components of ESD, with minimal progress made in social and economic areas (Millington and Fletcher 2008). Furthermore, fisheries management agencies continue to fail to effectively engage with the community; and there remains no baseline social/economic assessments of fisheries or regions (Millington and Fletcher 2008). This is particularly critical given the fact that it is 'apparent' that climate change will significantly impact Australian fisheries, and that "considerable social disruption would occur if key fisheries were affected" (Hobday, Poloczanska and Matear 2008, 14). Indeed, as do several others, these authors go on to suggest the need to investigate the socio-economic impacts of climate on Australian fisheries and aquaculture. While the research underpinning this book focuses on discourses of 'crisis' and 'sustainability', further demonstrating the currency of this work is the fact that the discourse of impending climate change may be critically viewed as the latest industry 'crisis' (Crist 2007).

A further purpose of this book is to provide emerging feminist scholars with a sound post structural theoretical framework for undertaking feminist research. Undertaking higher degree research can be daunting, and many students struggle with determining and articulating a solid framework for research planning, data collection and analysis. Indeed, it was my early experience that accessible and comprehensive material relating to reconciling theoretical and methodological tensions was difficult to locate and understand. As such, I hope that this book assists emerging scholars in both developing a greater understanding of these issues, and also provides an accessible example of how to apply this knowledge to a particular subject to advance the positioning of women.

Women in Fishing-The Issues

Historically, commercial fishing has been characterised by family businesses operating in the wild catch sector (Wilkinson 2004). Traditionally, fishing has been constructed as a singular, masculine occupation and, despite the reliance on labour from women in family businesses, women's contributions have not been visible (Lambeth et al 2002; Siason et al 2002). The application of a feminist informed Foucauldian theoretical approach highlights the importance of analysing the what, where, why and how of women's contributions within a specific historical context. Thus, exploring the current industry climate of fisheries in "crisis" (Kura et al 2004; *Fishing Crisis* 2006) and discourses of sustainability are central to this analysis.

International research suggests that women in fishing make vital contributions yet lack recognition within a masculine dominated industry (Thomspon 1985; Neis et al 2005; Siason et al 2002). In Australia, the small amount of research conducted with women in fishing also suggests that this is the case (Aslin, Webb and Fisher 2000; DNRE n.d.; Stella 1996, 1998). However, there has been no thorough exploration of these issues, or the implications of women's lack of stakeholder status in the current industry climate and the potential of women to enhance the future of the Australian commercial fishing industry.

Nonetheless, there is a growing body of international feminist literature on women in commercial fisheries. Nadel-Klein and Davis's (1988) recognition of women's extensive contributions to fishing globally created impetus for research to further explore these issues in the early 1990s, much of which was conducted in Norway, Iceland and Canada (e.g., Porter 1991; Davis and Nadel-Klein 1992; Thiessen, Davis and Jentoft 1992; Davis 1993). In an effort to alleviate the male bias in fisheries research, both *Anthropologica* (Binkley 1996; Hall-Arber 1996; Skaptadottir 1996; Szala-Meneok 1996) and *Women's Studies International Forum* (Binkley 2000, Davis 2000; Gerrard 2000; Nadel-Klein 2000; Neis 2000; Munk-Madsen 2000; Skaptadottir 2000) published special journal issues dedicated to women in fisheries and fishing communities which recognise women's extensive and diverse contributions, often in the context of fisheries in "crisis". The more recent publication of *Changing Tides: Gender, Fisheries and Globalization* (Neis et al 2005) continued to position women's extensive contributions and marginalised position within the broader context of a changing fishing industry, and explore the implications of these changes for women in fishing (Binkley 2005; Neis 2005; Neis and Maneschy 2005; MacDonald 2005, McCay 2005; Power and Harrison 2005; Skaptadottir and Proppe 2005).

The synergies between women in fishing and women in farming are worth noting and are further explored in Chapter 1. Over the past three decades, an extensive amount of research has emerged which focuses on women in farming. Recognising the patriarchal, masculine dominated nature of farming, this research has explored numerous issues including the visibility of women in farming, their contributions, their representation in industry decision making and transforming gendered industry identities both in Australia (Alston 1995b; 1998a, 2000, 2006; Bryant 1999, 2001; Davidson 2001; Dempsey 1992, 1994; Gibson, Baxter and Kingston 1990; Haslam-McKenzie 2003b; Liepins 2001; Pini 2002a, 2003b, 2004, 2005a, 2005b; Poiner 1990, 1994) and internationally (Brandth and Haugen 2000, 2005; Little and Morris 2005; Shortall 1999, 2000, 2003, 2004a, 2005; Whatmore 1991a, 1991b; Whatmore, Marsden and Lowe 1994).

However, studies investigating women's contribution to the Australian commercial fishing and the gendered nature of the industry are a more recent phenomenon, with several authors identifying the need for further research in this area (Aslin, Webb and Fisher 2000; MyriaD Consultants 2000; Stella 1996, 1998; Waitt and Hartig 2005; Smith and Sullivan 2001). Women's knowledge, gained through the experience of engaging in their daily activities, has also been identified as important for enhancing natural resource management and facilitating more sustainable agricultural industries (Agarwal 1992; Sachs 1996; Trauger 2004). This recognition occurs within a global context of increasing acknowledgment of the importance of women's contributions to more sustainable development and industries (Fourth World Conference on Women 1995; SCARM 1998, 2001). However, while there has been a sustained movement to increase the participation of women in farming, women in fishing have been less likely to be incorporated into this rhetoric.

Nonetheless, women in fishing in Australia have started to organise and activism by the Women's Industry Network Seafood Community (WINSC), a national network of women in the seafood industry in Australia formed in 1999, has been promoting women in the industry. At this time of this research, WINSC had also had representation on the Rural Women's Secretariat (RWS) in the Commonwealth Office for Women, and the Regional Women's Advisory Council (RWAC), a representative body located in the Commonwealth Department of Transport and Regional Services. In addition, WINSC has been instrumental in generating research to increase recognition of women in fishing, including facilitating a project *Fishing for Women: Understanding Women's Roles in the Fishing Industry* (Aslin, Fisher and Webb 2000) which was conducted by the Bureau of Rural Sciences (BRS), a subsidiary of the Commonwealth government department at that time, Agriculture, Fisheries and Forestry Australia (AFFA). Indeed, this report created the impetus for an application to be made to the Australian Research Council for this doctoral research to be undertaken.

In order to generate more critical debate on women in fishing, this book explores the nature of the dominant fishing industry discourse; women's involvement in industry decision making; and their contributions to their businesses (including family business decision making), families and communities. It then explores the implications for women in the current industry climate and their potential to enhance the future of the fishing industry, with a focus on the value of women's knowledge and sustainability. This exploration is achieved through research conducted with women in fishing in three case studies–the Western Australian Rock Lobster fishery, the Tasmanian Oyster fishery and *Cardea*, a pseudonym for an Australian east coast fishing village where women are involved in the Commonwealth Trawl Fishery and tuna fisheries. As with any feminist research, this book is unashamed about exposing inequalities with a view to challenging gendered assumptions that marginalise women in any social arena (Reinharz 1992), including the fishing industry.

My Story and This Research

A further critical aspect of the feminist nature of this research and my adoption of reflexivity-a position which rejects the notion of an impersonal objective researcher-suggests I disclose my biography as a researcher so my beliefs and social location may be considered in the context of the overall research (Harding 1987b). When this research was undertaken, I was in my mid thirties, and a white, middle class feminist with two teenage children. As such, it is difficult to know what is important to disclose about myself for it is not possible to articulate all of my values, knowledge and beliefs and the factors which account for these, so the following is only a brief overview of some aspects of my life. Ironically, this divergence between my knowledge of self and the readers reinforces the underpinning theoretical construction of knowledge in this book as partial, situated and located (Haraway 1988).

After leaving school with only four years of secondary education, I was successful in gaining a university place under a special admissions program and commenced an educational journey which has and will continue to mark my personal development. I have also had a long standing interest in and passionate commitment to women's issues. My honours thesis investigated the shifting sexual division of labour in restructuring industry, and I have worked in organisations and conducted consultancy work where empowering women through both research and action has been the central aim.

When I commenced my doctorate, I was living in a small coastal town seeking stimulating employment which provided the flexibility to fit in with my family responsibilities. In rural areas, these opportunities are scarce. One evening I happened to come across an advertisement for a research scholarship, and having completed an undergraduate degree in psychology and sociology and an honours year in sociology, met the required academic criteria. While I had no prior knowledge of, or experience in, the commercial fishing industry, I also lived in an area with a high profile fishing industry. The possibility of applying a feminist perspective to exploring this industry appealed to me, and I was privileged to be successful in gaining this opportunity which I undertook as a distance education student, a common form of study in Australia where students undertake educational programs away from their residential location.

My experience of the process of engaging in and undertaking a doctorate was a personal journey. Not only have my conceptual, analytical, research and writing skills advanced significantly, but my own subjectivity has been influenced by my enhanced understanding of, and affiliation with, feminism. My commitment to women's issues and passion for social justice issues has strengthened with the insights I have gained. It is with great delight that I now have the opportunity to publish this research, and my hope that this work enhances the position of industry women into the future.

A Feminist Post Structural Approach

Over the last two decades, a significant influence on research with women in agriculture has been the application of post structural theory. This has generated a more sophisticated understanding of the diversity amongst women, their agency and resistance, and the positioning of women in masculine industry discourses which moves beyond theorising "women" as an essentialist category inscribed as victims (Brandth and Haugen 1998; Grace and Lennie 1998; Leipins 1995, 1996, 1998, 2001).

While I felt there were significant benefits in applying Foucauldian post structuralism, I also became familiar with the feminist critique of Foucault (Bailey 1993; Cain 1993; Deveaux 1994; Grimshaw 1993; Ramazanoglu 1993a). Issues raised by these authors include: the notion that women's daily activities lack a material reality and instead are seen to derive only from discursive subject positions; the lack of attention given to internal processes and emotions in taking up and rejecting subject positions; and the difficulty in discussing collective political action for "women". For these reasons, I draw from both post structuralism and standpoint theory, and reconcile the tensions between these positions by employing the concept of "situated knowledge" (Haraway 1988).

A further benefit of applying a post structural theoretical approach is its emphasis on the need to explore power in specific local and historical contexts (Foucault 1980). As power is theorised by Foucault as everywhere (capillary) and *exercised,* and subjectivities are constituted in and by discourse, exploring power relations at a broad contextual/ structural level is required in order to understand what women do, and do not do, and why and how this occurs. This aligns with the contextual understanding that the fishing industry is operating under increasing global, local, environmental, economic and social pressures (Neis et al 2005).

Key Definitions-Language and Meaning

In order to aid understanding of the reader and as meaning cannot be guaranteed by any author (Weiner 1993), I now provide a brief explanation of how key terms are employed throughout this book. Theoretical concepts are explained in more detail in Chapter 2.

Discourse

The construct of discourse which is applied in this research is understood as:

> "...practices that systematically form the objects of which they speak...Discourses are not about objects; they do not identify objects, they constitute them and in the process of doing so conceal their own invention" (Foucault 1980, 49).

Family businesses

In this book, I employ Tonts and Black's (2002) conceptualisation of the family farm business.

> "...family farms are usually held to be businesses that are owned and operated by a family. The decisions about farm management are made

> within the family unit, which also provides the business with much of the necessary land, capital and labour" (Tonts and Black 2002, 3).

However, in doing so, I transpose the key elements of this definition to family businesses in the fishing industry. As such, I transpose family farming business with family fishing business; decisions about farm management with decisions about management of the fishing business; and land and agricultural capital with fishing capital including boats, fishing gear, quota, leases and licenses.

Fish, fisheries and the industry

The definition of "fish" as "…living aquatic vertebrate and invertebrate organisms…" (FRDC 2001, 6) includes fish and shellfish harvested in the aquaculture and wild catch sectors of the Australian commercial fishing industry. A "fishery" refers to one of the approximately 150 sectors within the industry at the time of this research which are delineated by species and/or geographic area. While "the fishing industry" is often commonly understood as referring to the act of harvesting fish within these fisheries, an official definition recognises "the fishing industry" as comprising of a broad range of activities.

> "The fishing industry includes any industry or activity conducted in or from Australia concerned with taking, culturing, processing, preserving, storing, transporting, marketing or selling fish or fish products" (FRDC 2001, 6).

While I employ this definition, I further expand this term and use it to refer to those individuals who undertake these tasks, including women, men and children involved in fishing.

Fisheries management agencies

"Fisheries management agencies" refer to the State or Commonwealth government department responsible for implementing policy within specific fisheries. In Australia, the Australian Fisheries Management Authority (AFMA) oversee Commonwealth fisheries, including the Eastern Tuna and Billfish Fishery and the Commonwealth Trawl Fishery which both operate in Cardea. State fisheries management agencies relevant to this research include Western Australian Fisheries, who are responsible for managing the Western Australian Rock Lobster fishery and the Department of Primary Industries, Water and Environment (DPIWE), who are responsible for managing the Tasmanian oyster fishery.

Industry organisations

The term "industry organisation" is used to refer to any organisation, association or body which is comprised primarily of commercial fishing industry stakeholders. One example of these at a local level are the Professional Fishermen's Associations in the Western Australian Rock Lobster fishery. An example at a state level is the peak organisation, the Tasmanian Seafood Industry Council, which represents all commercial fishing operators in the state of Tasmania, and at the time of the research was called the Tasmanian Fishing Industry Council (TFIC). I employ the term "industry decision making" to refer to both these decision making structures, and the decision making processes which occur within these structures.

Participant

I employ the term "participant" when referring to both women and men who were involved in this research, the latter typically being representatives of industry organisations and fisheries management agencies.

Power/ knowledge and truth

Power is theorised as capillary, multiple, productive and as constituted, embodied and exercised in all relations and subjectivities, and as inseparable from resistance. Furthermore, power/ knowledge is conceptualised as unified and singular, and as the fundamental component of what defines truth in the dominant discourse.

Subjectivities, subject positions and positioning

I employ the term "subjectivity" to mean "…the conditions of being subjected to frameworks of regulation, knowledge and discourse and constructing subjectivity in the process" (Skeggs 1997, 12). "Subject positions" refer to categories constructed as part of wider discourses as an effect of discourse and structures which "…investigate the specificities of how women become particular subjects, especially respectable subjects" (Skeggs 1997, 12). I use the concept of "positioning" to refer to how individuals are defined in, and by, particular subject positions.

Sustainability and neo liberalism

The language of "sustainability" is prevalent in industry and fisheries management discourse. In this research, I have not conceptualised "sustainability" as a theoretical construct, but as a linguistic, discursive site of power. As such, I understand "sustainability" and "a sustainable industry" not as an apolitical, concrete, achievable state of being or end result, but as a term which is shot through with power and which refers to a variety of ideas with differing meanings (depending on who is speaking and who is hearing) relating to the need for all or some things to continue to survive and how this may occur. Within this post structural framework, what becomes pertinent is how the language of sustainability is employed, for what purpose, and the implications of the way it is used.

At the time of this research, the language of sustainability was both dominant, and used in a neo liberalist state culture. In this research, I employ the term neoliberalism to refer to

> "...an all-embracing economic and political ideology that advocates the supremacy of the market over any alternative social arrangement....The market, based on freedom of choice and respect for private property and individual rights, and underpinned by competition...is seen as the ideal and optimal vehicle for the realization of human ends (Mhone 2005, para 4)".

Women and men in fishing

In this book, I employ the terms women and men "in fishing" and "in farming" to refer to women and men involved in family fishing and farming businesses. I employ the terms "men in fishing" and "women in fishing" to refer to individuals involved in family fishing businesses within the fishing industry. In fishing, typically these women are married to men who harvest fish. The terms "fisherman" and "fisherman's wife" are used when referring to the prevalent industry subject positions for men and women in fishing. In order to emphasise the constructed nature of these subject positions and other terms, I use inverted commas, and in order to emphasis the masculine occupational privileging in particular contexts within the fishing industry, I use round brackets as a technique to draw attention to the relationship between "man" and "fisher". In the literature review, my use of the term "fisher" reflects the use of this term by particular authors.

Women's contributions

When writing of the "contributions" of women in fishing, I refer to the range of activities in which women in fishing and farming are engaged. Women in these industries make extensive contributions to their businesses, industries, families and communities. I conceptualise these contributions as holistic, as these activities are not easily separated into distinct spheres of business, industry, family and community. As such, women's contributions are intertwined, and may be social, cultural, economic, and/ or political.

Addressing the Issues

Masculine industry discourses

Women's lack of visibility is a concept that lies at the core of feminism and, given the lack of visibility of women in the Australian commercial fishing industry, is central to this research. In this book, in addition to drawing from overseas literature on women in fishing, I also incorporate themes drawn from research on women in farming for three main reasons. Firstly, similarities have been observed between women in fishing and farming including the historically male dominated nature of these industries, the prevalence of family units and patrilineal inheritance practices (Aslin, Webb and Fisher 2000). Secondly, there is often a lack of distinction between the two industries. Statistics on fishing and farming are often aggregated in Australian Bureau of Statistics (ABS) labour force data (e.g., Herreria et al 2004) and the industries are linked by their primary industry status in the Commonwealth government's Department of Agriculture, Fisheries and Forestry, which at the time of this research was called Agriculture, Fisheries and Forestry Australia (AFFA). Thirdly, this research's exploration of women in fishing in Australia replicates the initial stages of a similar process that occurred for women in farming some 30 years ago. As such, incorporating themes which emerge from research on women in farming enables a more sophisticated analysis to be conducted of the experiences and positioning of women in fishing.

Three central themes emerge in previous research on women in fishing and farming. Firstly, women's lack of visibility, including their poor representation in industry decision making and the masculine nature of "fisherman" and "farmer", are challenged. Secondly, women's diverse and extensive contributions, including their paid and unpaid labour, are recognised. Thirdly, barriers and opportunities that impact on women's ability to participate in the industry are examined, and strategies are identified to facilitate the increased representation and participation of women in fishing and farming. These themes emerge within a broader context which highlights the continual transition and shaping of industries by local, global, social, economic and environmental forces intersected by government policy. This context is often referred to as the "crisis" in farming (Bourke and Lockie 2001; Rickson 1997) and fishing (Kura et al 2004; *Fishing Crisis* 2006).

The current industry climate

At any given time, the Australian commercial fishing industry is affected by economic, social and historical factors intersected by government policy. Currently, the quest to achieve more sustainable fisheries is occurring in the context of a global fisheries "crisis" (Kura et al 2004; *Fishing crisis* 2006). In Australia, this is resulting in the emergence of aquaculture as a major growth sector (FRDC 2000; FRDC 2001; NAC 2001) and wild catch fisheries are operating under increasingly stringent fisheries management regulations (Caton and McLoughlin 2000). Under these pressures, the number of smaller family businesses in the industry are decreasing and the number of larger, more corporate businesses are increasing. This displacement of family businesses is occurring in both farming (Burch et al 1998) and fishing (Aslin, Webb and Fisher 2000; Wilkinson 2004).

Exploring the implications of the current positioning of women within this broader context is essential to ensure that industry women are not (further) marginalised in processes of change. Overseas research suggests that the current patriarchal bio-economic model employed in fisheries management is formalising masculine privileging in the inequitable allocation of resources (Munk-Madsen 1998a) and is failing to adequately consider social issues in restructuring fisheries (Davis 2000; Davis and Gerrard 2000; Neis 2000; Skaptadottir 2000). In Australia, it has been reported that women are concerned about the security of their fishing rights and question whether they will be able to pass these rights onto their children (Aslin, Webb and Fisher 2000; MyriaD Consultants 2000). Elsewhere, based on this doctoral research, I have argued that women's exclusion from industry decision making, deriving from their marginalised ownership status, means that women in family fishing businesses are not recognised in the process of the allocation of the rights to resources and not adequately considered with regard to the social impacts of management decisions (Dowling 2006a).

Research questions and objectives

This research builds on the small amount of work conducted on women in fishing in Australia by addressing the five following questions that emerge as a result of the literature review. How visible are women in the industry? How do women contribute to the industry, their businesses, families and communities? What are the barriers to and opportunities for women's greater participation in the fishing industry? What are the implications of women's industry positioning in the current climate? What outcomes would women like to see from this research?

Reflecting that the feminist agenda is about change, I also articulate the objectives of this research, where I aim/ed to: make visible and increase awareness of the contributions of women in family fishing businesses; give women "voice" and expose their subjugated knowledge; identify, recognise and value the lived experiences of women in order to increase the opportunities and potential for women's empowerment; expose the barriers to and opportunities for women's contributions greater recognition and participation in the fishing industry; enhance women's capacity to contribute in the context of an industry in transition; raise awareness of women as stakeholders based on principles of social justice; provide a platform for facilitating broader political action that increases recognition of women as industry stakeholders and aspires to more equitable industry relations; and contribute to scholarship in the fields of feminism, social science, and environment and natural resource management.

The Book Structure

In order to explore women's lack of visibility and their current and potential contributions to the Australian commercial fishing industry in the current industry climate, this book is presented in a further eight sections–seven chapters and a Conclusion. Chapter 1 reviews the literature; Chapters 2 and 3 articulate the theory and methodology employed; Chapters 4, 5, 6 and 7 present the findings of this research with women in fishing; and the Conclusion provides a comprehensive summary of the key findings and their implications.

Chapter 1

Chapter 1 is a review of the literature that draws from scholarship and research on both women in fishing and farming. This literature review explores the key themes of women's lack of visibility; their contributions to their businesses, families and communities; their involvement in business and industry decision making; and the barriers to, and opportunities for, women's greater recognition and participation. It also explores discourses of sustainability in the context of fishing and farming in "crisis", and gendered management practices in the current climate.

Chapter 2

In Chapter 2, I present the feminist theory employed in this research. As previous research has suggested, there is a need to go beyond documenting women's "roles" to explore how women are positioned in discourses and the implications of this positioning (Davis and Nadel-Klein 1992). This requires a move away from essentialist structural understandings where oppression is "fixed onto women" to an approach which recognises diversity and reinscribes women with power and the potential for agency, resistance and empowerment.

The theoretical position adopted in this book, drawing from Foucauldian informed feminist post structuralism and standpoint theory, is then articulated. This follows the tradition of other feminist writers who argue for the benefits of reconciling the ontological tension between the constitution of reality from experience in standpoint theory, and from discourse in post structuralism, by locating knowledge as historical, situated and partial (e.g. Haraway 1988; Hirschman 2004; Weeks 2004).

Adopting this approach exposes the inherent instability in universal Truths and recognises diversity amongst women and the intersecting nature of socially constructed locations/positions such as sex/ gender and race/ ethnicity. It allows for an analysis of women's material experiences and discourses of power/ knowledge which are constituted in and through women's fluid subjectivities. It facilitates the exploration of women's subjectivities, resistance and empowerment at both an individual and group level, and has the potential to expose subjugated knowledge.

As meaning cannot be guaranteed by an author's intent, this is followed by a more detailed exploration of my interpretation and application of the key post structural concepts critical to understanding the positioning of women in the Australian commercial fishing industry. This chapter concludes with a section on social positions where I justify my decision to focus on sex/gender and race/ethnicity as the primary social locations under investigation, and explain my omission of class.

Chapter 3

Chapter 3 overviews the feminist methodology employed in this research and explains the research process-the "doing" of feminism (Kelly, Burton and Regan 1994, 28). Initially I discuss epistemological and methodological issues relating to feminist research. The principles which inform this feminist research are articulated, the criteria for evaluating feminist research discussed and feminist concepts of validity and generalisation explored. Each section also explains how these concepts and principles were applied throughout this research. A detailed explanation of, and justification for, the research design and process is then provided.

The research is based on pragmatic, feminist, philosophical underpinnings and is informed by Pini's (1998) articulation of the five recurring common themes in feminist research: a focus on gender, value given to women's experience's and knowledge, rejection of the separation between subject and object, an emphasis on consciousness raising and an emphasis on political change.

Chapter 4

Chapter 4, A Masculine Industry Discourse, explores the gendered power relations in the dominant industry discourse and how women are positioned in this discourse. I do this by deconstructing both clearly gendered language such as "fisherman", and seemingly apolitical language such as the meaning commonly ascribed to "the industry", as a means of exposing gendered power relations. In this chapter I name the central industry subject positions of "fisherman" and "fisherman's wife", and acknowledge the discursively relegated/ adopted activities associated with these positions.

I then explore the gendered power relations in women's positioning in family businesses in relation to their legal ownership of businesses and capital, and explore women's marginalisation through patrilineal inheritance practices and the implications of these practices. The subject positions of "fisherman" and "fisherman's wife" are then scrutinised in more detail and the discursive normalising and regulatory mechanisms which exclude women from being occupationally privileged are explored. I conclude by discussing the implications of this occupational privileging and the positing of women within the dominant discourse.

Chapter 5

The purpose of Chapter 5, Women in Industry Decision Making, is to examine and explain women's visibility, or lack of visibility, and their contributions in industry decision making. Women in farming are marginalised in industry decision making as a result of the power conferred through patrilineal inheritance practices (Shortall 1999), structural factors such as masculine dominated organisational practices and processes, and personal reasons such as a lack of time (Alston 2000). Research suggests that women in fishing in Australia are marginalised in decision making in similar ways (McKay, Dimopoulous and Baker, n.d).

This chapter explores women's representation in decision making bodies relevant to the Australian commercial fishing industry. Exploring the meanings associated with the subject position of "fisherman" and "fisherman's wife", it explores the positioning of industry women in relation to decision making. I then discuss discursive regulatory mechanisms which operate to perpetuate this positioning of industry women including structural barriers such as organisational processes and practices. Analysis is also provided on how women resist these barriers and the consequences for women who breach the dominant discursive positioning of women. Strategies which may enhance the potential for women to contribute are then explored, including women's collective action and activism.

Exploring women's contributions in this context reveals the implications of women's lack of occupational privileging. This is particularly pertinent as the industry comes under increasing pressure to change. These implications are further explored in Chapter 7, Women in an Industry in "Crisis".

Chapter 6

Despite women's lack of recognition, previous research has demonstrated that women in farming and fishing make substantial contributions to their businesses, industry, families and communities. The small amount of research on women in fishing in Australia indicates these findings are replicated (Aslin, Webb and Fisher 2000; Stella 1998). Having established that women are not privileged with occupational (and so stakeholder) status, and are excluded from industry decision making, in Chapter 6- Women's Contributions, I provide a comprehensive overview of the contributions women in fishing make to their businesses, families and communities. These contributions are explored in a holistic manner and include women's labour, economic, social and emotional contributions. Reflecting my application of feminist post structural theory, I commence this chapter with an analysis of how women in fishing are discursively relegated into and take up (or resist) the daily activities discursively associated with the subject position of "fisherman's wife". I conclude this chapter with a discussion on the knowledge/ power implications of women's engagement in their daily activities.

Chapter 7

Chapter 7 explores women's invisibility, their contributions and lack of recognition within the broader changing industry context characterised by increasing regulation and management intervention. I commence with a critical deconstruction of the language employed by management with a focus on the language of "sustainability" and an assessment of how adequately, within the current neo liberalist, masculine dominated, bio economic approach (Holm 1996; Munk-Madsen 1998a) fisheries management integrate the "social" in sustainable. This analysis exposes the political intent in an apparently apolitical analysis. The implications for women within this context are then examined. I then proceed to explore the potential for women's knowledge/ power which derives from their holistic activities as "fishermen's wives" to contribute to enhance the industry, and the likelihood that this will occur, particularly given their current exclusion from industry decision making.

Conclusion

The Conclusion provides a comprehensive summary discussion on critical issues and implications pertaining to the visibility of women in fishing, their contributions, and the barriers and opportunities for their greater industry participation in the current industry climate.

CHAPTER ONE

LITERATURE REVIEW

Introduction

The paucity of literature on women's involvement in the Australian commercial fishing industry highlights women's lack of visibility, and the lack of recognition for their contributions. This chapter draws from literature on women in fishing internationally and the small amount of literature on women in fishing in Australia. It also draws on literature on women in farming, particularly in the Australian context. Incorporating themes that emerge from this literature on women in agriculture not only enables a more sophisticated analysis of women's involvement in commercial fishing, but also provides critical information on the Australian context.

This chapter reviews the literature focusing on the visibility and contributions of women in fishing and farming, and examines the three central themes that emerge. Firstly, women's lack of visibility and the gendered meanings of "fisherman" and "farmer" are explored. This includes a review of women's representation in official statistics and the media. Secondly, the contributions that women make in terms of paid and unpaid labour and input into decision making, both at a family business and industry level, are articulated. Thirdly, barriers to and opportunities for the greater participation and recognition of women in fishing and farming are discussed throughout the chapter.

This is followed by an examination of the implications for women in fishing and farming in the broader context in which these themes emerge-the continual transition and shaping of industries by global, local, social, economic and environmental issues, which are intersected by government policy. A central aspect of the broader context is the "crises" in fishing and farming and the policy responses employed to address these crises in the quest for "sustainable" fishing and agricultural industries. This literature review suggests that, while women make vital contributions in the fishing and agricultural industries, they lack recognition and are at risk of being further marginalised within the current context of management and policy responses to these industries in "crisis".

Visibility of Women in Fishing and Farming

Masculine constructions of fishing and farming

The "fishing industry" is an inherently masculine construct typified by heroic fisher*men* braving danger on the sea and hunting from Mother Nature. In commercial fisheries, this is reflected in imagery and textual representations of deep sea fishermen away at sea for long periods of time who "...rely on large powerful boats to penetrate deep into the ocean..." (Okraku 1975, 328). The gendered term "fisherman" reflects men's privileging with occupational status (Dixon, Lowery and Sabella 1984; Yodanis 2000). Women, as "fishermen's wives", are excluded from occupational status and from fishing, and rendered passive and secondary (Dixon, Lowery and Sabella 1984). Similarly, there is little historical evidence of the participation of women in farming with the gendered term "farmer" traditionally referring to men involved in the occupation of farming. Like women in fishing, women in farming have held a secondary status as "farmer's wives" or "farm wives" (Alston 1995b; Hogan 1994).

However, women in farming have made and continue to make extensive contributions directly and indirectly to farm productivity and rural communities (Alston 1990, 1991, 1994a; Gibson, Baxter and Kingston 1990; Jennings and Stehlik 2000). Despite the masculine construction of the agricultural industry, farming has been recognised as much more than an individual occupation, and instead as dependent on collective family labour which historically has been supplied by women and children (Fink 1990). Similarly in fishing, essential fishing activities in which women are actively engaged such as business administration, subsistence activities and post harvest activities are not recognised as aspects of the occupation of fishing or in constructions of the fishing industry (Lambeth et al 2002; Shariff 2002; Siason et al 2002). Yet, women's extensive, diverse and essential contributions to the sustenance of global fisheries render fishing "...an occupation peculiarly dependent on the work of women" (Thompson, Wailey and Lummis 1983, 167).

The narrow masculine construction of these industries is one factor in explaining the lack of recognition of women as industry stakeholders.

These masculine constructions have implications for the exclusion of women from industry decision making processes and policy development in both the fishing industry (Lambeth et al 2002) and the agricultural industry (Board 1998). Furthermore, women in fishing are marginalised in the allocation of the rights to resources in the current process of privatising resources (Dowling 2006a; Munk-Madsen 1998a).

Media constructions of women in fishing and farming

Exploring media representations of women in fishing and farming provides an insight into the way women are constructed and positioned. Research has explored the representation and positioning of women in farming through both content analyses (Bell and Pandey 1989; Sheridan 1994) and post structural analyses (Brandth and Haugen 1998; Liepins 1996).

Bell and Pandey (1989) conducted a content analysis of advertisements from *The Land*, a major Australian rural newspaper, and conclude that "farming was characterised as men's work and as almost entirely dependent on their labour….the notion of farmer is essentially masculine" (45). In contrast, women are portrayed as secondary to men, as sex objects or "housebound wives and mothers" (Bell and Pandey 1989, 45). Sheridan (1994) conducted a follow up review of advertisements in *The Land* which support these findings. She argues that the representation and lack of women in the media acts to perpetuate the invisibility of women in agriculture, the "man on the land" stereotype and the existing male hegemony (Sheridan 1994). Brandth and Haugen (1998) apply a post structural approach in their analysis of women in the Norwegian forestry press, and similarly conclude that "…the gender issue is marginal, both in terms of space and influence" (Brandth and Haugen 1998, 440). While the increasing construction of women as "experts" holds some potential for altering the masculine gendering in the industry, women are constructed as "different" to men in relation to forestry work, which is characterised as heavy, dirty and dangerous and connected to dominant notions of rural masculinity (Brandth and Haugen 1998).

Alternate discursive texts, such as women's network publications, do enable women's subjugated knowledge to be exposed (Brandth and Haugen 1998). In the Australian commercial fishing industry, one such site is the Women's Industry Network Seafood Community (WINSC). Since the early 2000s, there have also been several media articles in the popular Australian press profiling women in the fishing industry (eg Gorton 2001; Voorhaar 2000). However, while these articles highlight women in fishing, they tend to focus on the activities undertaken by high profile women in fishing and do not include their views or opinions on current industry issues. While Alston (2000) argues that media constructions of women in farming in the mainstream rural press reinforce and perpetuate "an insidious sexism" in the industry (220), there is an absence of both Australian and international research on media representations of women in fishing and their positioning in industry discourses.

Statistics on women in fishing and farming

Historically, women in farming have not been visible in industry statistics (Alston 1995b; Williams 1992) and more information is available on livestock production than on women's industry involvement (Women in Agriculture International Conference Committee 1995). Statistical collection techniques in the agriculture and fishing industries are gendered and underestimate or exclude women's involvement. In farming, data is often collected autonomously by different agencies and is difficult and expensive to locate (Haslam-McKenzie 2003a). Statistics are characterised by a lack of gender segregation (Haslam-McKenzie 2003b), response categories are often inappropriately labelled, and there tends to be insufficient categories to accurately reflect women's situation (Shortall 2003).

Similarly in the fishing industry, statistics are difficult to locate, are often not publicly available, do not reflect women's unpaid labour and tend to be gender aggregated (Lambeth et al 2002; Siason et al 2002). For example, the majority of Australian fishing employment statistics are aggregated with agriculture and forestry statistics (Lambeth et al 2002). At the time of the research, the Fisheries Research and Development Corporation (FRDC 2001) indicated that more than 110, 000 Australians were employed in the seafood industry–28, 000 in commercial fishing; 60, 000–70, 000 in processing, marketing and sales; and 10, 000 in aquaculture. However, no further information is provided on these statistics and it is likely that these figures only include women who are employed in remunerated positions in the fishing industry. Industry statistics continue to be aggregated and recognised as likely to under report industry employment (DAFF 2008). The small amount of available statistical information which incorporates women in fishing reflects women's marginalised industry status. For example, from a total of 46 participants from the wild catch fishing sector undertaking FarmBis training in 2000-2001, eight, or 17%, were women (PISC 2003).

The statistical exclusion of women's unpaid labour in farming has been attributed to the masculine construction of "work" (Alston 1995b). Farm work has been conceptualised as being contingent on productivity and payment, which equates with the labour of "farmer", and there has been a lack of recognition of the collective labour required to support farming businesses (Alston 1995b). As with fishing, the construction of farm "work" as paid and productive discounts the productive and unpaid labour of many farm women, renders women's work invisible and is an inadequate construction for analysing farm women's labour. In 1999, in an effort to rectify the paucity of statistics available on women in agriculture, the Australian Bureau of Agricultural and Resource Economics (ABARE) published *Farmers at Work: The Gender Division* (Garnaut, Rasheed and Rodriguez 1999). While this survey acknowledges the lack of recognition for women's contributions to their farming businesses, Alston (1998b) argues that the report continues to underestimate women's contributions due to the inadequate construction of "work" employed in the study. At the time of this research, no such survey had been undertaken for women in the Australian fishing industry.

Feminists and economists challenge the Keynesian economic approach which does not include housework, unpaid labour and volunteer work in the Gross Domestic Product (GDP) (e.g., Waring 1998). Important from a social justice perspective (Teather 1998), quantifying women's labour was the task of *Missed Opportunities: Harnessing the Power of Women in Australian Agriculture* (RIRDC/ DPIE 1998, 1998a). This expanded definition of work included paid on-farm work, paid off-farm work, unpaid on-farm work for commercial production, unpaid on-farm work for household production, including household activities not imputed in the National Accounts, and unpaid work in volunteer activities and community activities (RIRDC/ DPIE 1998). This survey reports that women's on and off farm work represents 48% of real farm income and women generate over 80% of off-farm income (RIRDC/ DPIE 1998). A follow up report, *Revisiting Missed Opportunities-Growing Women in Agriculture* (Sheridan and Haslam-McKenzie 2009) indicates that women's representation in decision making remains disproportionately low; women continue to lack recognition; and that women continue to contribute significant amounts of labour to their enterprise, communities and families. While no quantifications have been made of women's labour in the Australian commercial fishing industry, where the contributions of women in fishing overseas have been quantified, it is significant. For example, in Lao it is estimated that women's paid and unpaid labour represents one half of the 3% of GDP contributed by fisheries (Siason et al 2002).

Some authors note an increasing prevalence of women in farming in statistics (James 1989; Sheridan 1994). For example, Sheridan (1994) analyses the Australian Bureau of Statistics (ABS) Labour Force agricultural sector statistics from 1978 to 1992. She observes that while there is negative growth for male employers and self employed workers, there is a 43.5% increase in female employers and a 24.8% increase for female self employed workers. The increasing representation of women in agricultural statistics has been attributed to the sharing of income and ownership by male farmers to ensure maximum taxation benefits (James 1989) and the increasing number of women who self-identify as "farmers" as opposed to "housewives" (Sheridan 1994). Indeed, some overseas authors also find there is an increasing propensity for women to self identify as "farmers" (eg, Trauger 2004). This is consistent with Bryant's (1999) assertion that there is a "...chipping away of traditionality in occupational identities in agriculture..." (256). Due to the lack of available statistical data, it is not possible to comment on whether women in fishing are increasingly likely to identify themselves as fishers. However, in Australia at least, this is less likely to be the case, as the more established women in agriculture movement, which has focused primarily on women in farming, may be seen as influencing the way women in farming identify themselves.

Contributions of Women in Fishing and Farming

Women's labour contributions to fishing are diverse and influenced by specific economic, historical, social and cultural factors (Siason et al 2002; Thompson 1985). This diversity aligns with the feminist epistemological premise that knowledge is situated and partial (Haraway 1988) and provides a caution against making essentialist generalisations which fail to reflect specific contexts. In this section, I discuss literature relating to the historical contributions of women in fishing and overview the contributions of women in fishing and farming with a focus on the sexual division of labour. This provides a context for a latter section in this literature review, Implications for Women in the Current Industry Climate, where I explore the implications for women in the broader context of industries in "crisis" and the policy responses in the quest for more sustainable fishing and agricultural industries.

Historical contributions of women in fishing

There is no academic research on the historical contributions of women in the Australian commercial fishing industry, but anecdotal information suggests that historically women in fishing in Australia have made extensive contributions to their fishing businesses (Lambeth et al 2002). We do know that Anglo-Saxon women have worked as deckhands, cooks and processors on commercial boats in the Northern Prawn fishery since the 1970s (Robins and Sachse 1994). While Aboriginal women and men have and continue to engage in traditional fishing practices (FRDC 2000), we also know that Aboriginal women were employed as divers in the Australian pearl industry in the early 1900s (Lambeth et al 2002). However, there is no academic research which further explores the involvement of Aboriginal women in commercial fishing in Australia.

More information is available on the historical contributions of women in fishing overseas. For example, early records show that in 1710 women from Salem managed the books and business while their husbands were at sea (Ulrich 1982 cited in Kaplan 1988) and in the 19th century, women and children regularly accompanied their Captain husbands on whaling expeditions and assisted in navigating and commanding the ship (Whiting and Hough 1953). For the past 200 years in Scottish fisheries, women-both wives and daughters-have worked in both private and public spaces (Nadel-Klein 2000).

> "…fishermen's wives and daughters performed a wide range of tasks….digging mussels to use as bait in the hand lining; shelling the mussels and baiting the lines; keeping the men's feet dry before a voyage by carrying them across the tide line to the boats; helping to haul the boats in at low tide; and curing and selling fish. This list does not include other indispensable activities, such as the endless knitting of warm clothes and underwear, the constant scrubbing and cleaning to maintain household respectability, and, of course, managing the household and raising the children while the men were out to sea" (Nadel-Klein 2000, 363).

1901 census figures for Grand Bank, Newfoundland, indicate that, from a total population of 1170 people, 262 men and 252 women were involved in the "catching and curing of fish" (Porter 1988). In 1921, the last year figures were collected, 359 men and 304 women from a total population of 1605 people were involved in this occupation (Porter 1988). Davis (1983, 2000) also describes the historical contributions of women in this geographical region and acknowledges the "expressive" contribution of women to fishing families and communities.

> "Women were the anchors of emotional and moral strength in their families and communities. It was the task of women to worry over the safety and welfare of their families. Women's worry was a form of women's work and tied them into the fishing ethos and occupational identity in an active and valued fashion" (Davis 2000, 347).

Contributions of Women in Fishing and Farming in Australia

The historical material on women in fishing indicates the integrated, holistic nature of their contributions across the spheres of business, family and community. As in farming (Alston 1994a; 1995b), several authors note the complementary nature of men's and women's labour in small fishing businesses where the family unit is also the economic unit (Davis 2000; Nadel-Klein 2000; Porter 1983), and the intrinsic interdependence between business and family relationships (Binkley 2000). Furthermore, women may gain a sense of identity and personal fulfilment from their involvement in their family farm businesses (Alston 1994a; 1995b).

While we have limited knowledge of women in fishing in Australia, we know far more about women in farming in Australia. This material is important as it provides some insight into women in family based businesses in primary industry in the Australian context. We know that women in farming make significant contributions to their businesses, families and communities (Alston 1994a, 1995a, 1995b; Gibson, Baxter and Kingston 1990; James 1989; Jennings and Stehlik 2000; O'Hagan, Alston and Spriggs 2003). Women in farming are responsible for: household labour including subsistence tasks which free up cash for investing into farm businesses; significant unpaid work such as book keeping and business management, driving machinery, working with livestock and "go-fering"; undertaking off farm paid work that often supplements the family income; and providing substantial voluntary labour in local communities (Alston 1995a, 1995b).

While anecdotal evidence indicates the significant but unacknowledged contributions of women in the Australian commercial fishing industry (Lambeth et al 2002; Ward 2004), since the mid 1990s, a small amount of research has been conducted on women in fishing in Australia that acknowledges their extensive contributions (Aslin, Webb and Fisher 2000; McKay, Dimopoulous and Baker n.d.; MyriaD Consultants n.d. 2000; DNRE n.d.; Stella 1996, 1998; Smith and Sullivan 2001). Aslin, Webb and Fisher (2000) find that more than 50% of women in fishing participate in bookkeeping, business administration and attend meetings. Stella (1998), in her unpublished PhD thesis *Trawling Deeper Seas: the Gendered Production of Seafood in Western Australia*, concludes that while women contribute significantly to the industry through a range of unpaid and paid labours, their work is rendered invisible through the masculine domination in the industry, broader societal gender relations and the rhetoric of the family and family business (Stella 1998).

The sexual division of labour and women in fishing

A major theme which emerges in the literature is the manner in which the contributions of women are mediated by the sexual division of labour. In contrast to women in farming, the sexual division of labour in fishing often reflects a land/ sea dichotomy where women's substantial contributions are land based, while men are responsible for sea based tasks. Indeed, "going to sea" in many fisheries and communities is almost exclusively a male activity accessed only by "fishermen" (Davis 2000; Gerrard 2000; Gronbech 2000; Nadel-Klein 2000; Porter 1983; Ram 1991; Rana and Choo 2002; Siason et al 2002; Skaptadottir 2000; Thompson 1985; Thompson, Wailey and Lummis 1983; Yater 1982).

> "Many observers of Newfoundland fishing communities have commented on the traditional, highly gendered nature of adaptation to fishing as a way of life, where men control and dominate the occupations directly involved with use of the sea. A land/sea division of labour defines the respective natures of men's and women's work, status, and power in their communities and households. A single occupational ethos dominates the local mental ecology" (Davis 2000, 344).

This is consistent with the only academic research on women in fishing in Australia by Stella (1996, 1998), who reports a rigid sexual division of labour on Western Australian prawn and scallop trawlers, and finds that this sexual division of labour also exists in fishers' households and in a processing plant in a Western Australian fishing community. More specifically, she finds that women in family based businesses are responsible for many of the land based activities for the family fishing business, household tasks and caring for children, while women who fish on boats cook, clean and help with deckhand work (Stella 1998).

There is a much greater volume of international research that explores the sexual division of labour and, despite cultural differences, this gendered labour pattern is often evident in fishing communities across the world (Nadel-Klein and Davis 1988). Activities in which women in fishing are engaged include harvesting and preserving fish for domestic purposes; fish selling and marketing; unpaid contributions to their businesses including bookwork, administration, the interpretation of fisheries regulations, ordering gear, cooking for the crew, boat and gear maintenance and processing; aquaculture tasks including pond cleaning, sterilising, feeding, harvesting and selling; and a range of other supporting activities (Allison, Jacobs and Porter 1989; Davis 1986; Howell 2002; Thiessen, Davis and Jentoft 1992; Lambeth et al 2002; Rubinoff 1999; Siason et al 2002; Thompson 1985; Thompson, Wailey and Lummis 1983; Yater 1982). In addition, women contribute financially to supplement household incomes, often through employment in the processing sector (Allison, Jacobs and Porter 1989; Davis 1986; Lambeth et al 2002; Porter 1991; Rana and Choo 2002; Siason et al 2002; Yater 1982); maintain domestic and family responsibilities (Allison, Jacobs and Porter 1989; Porter 1991; Rubinoff 1999; Siason et al 2002); maintain personal and community identity (Davis 1983, 1986); and are involved in community volunteer work (Davis 1986; Gerrard 2000; Porter 1991).

As the sexual division of labour transfers into different contexts, there are implications for women when they venture into "men's" territory. Women who crew on boats in Western commercial fisheries are often required to prove they are at least as capable as men (Kaplan 1988; Stella 1996, 1998). Even when women crew on boats, they may be expected to undertake "women's work" on board the boat. For example, women who crew on boats in Western commercial fisheries are often expected to cook (Kaplan 1988; Stella 1996, 1998). Women who fish may also be expected to engage in sexual relations and may be subject to sexual harassment (Kaplan 1988; Stella 1998). Some women also make conscious decisions to ignore sexual harassment at the expense of their physical safety (Howell 2002) and to employ sexuality to their utmost benefit in order to enter and remain in the fishing industry (Allison, Jacobs and Porter 1989; Stella 1998).

Women's exclusion from sea based activities is perpetuated through pervasive patriarchal industry relations. These gendered power relations are evident in cultural practices, beliefs and taboos in both commercial fisheries and non Western fisheries (Thompson 1985; Thompson, Wailey and Lummis 1983; Williams 2002). In commercial fisheries, women's exclusion is perpetuated through beliefs that women lack the necessary physical capacity to fish and the superstition that women on boats bring bad luck (Kaplan 1988; Stella 1998). Further barriers which exclude women from this sphere include: women's limited practical skills and knowledge of fishing technology which is attributable to their lack of participation (Siason et al 2002); a working environment which includes wet and dangerous conditions (Alison, Jacobs and Porter 1989); the absence of toilets on vessels (Aslin, Webb and Fisher 2000; Siason et al 2002); isolation from family and friends (including ethnic isolation where Anglo Saxon fishing women do not share in the dominant fishing culture); and a lack of acceptance by the female partners of other male crew members (Kaplan 1988).

While it is critical to draw attention to gendered practices which may be described as reflecting a sexual division of labour in fisheries, applying the concept of a distinct and definite sexual division of labour can be viewed as essentialist. Instead, applying a feminist post structural approach that recognises women's diversity, agency and the influence of specific historical, social and cultural contexts, requires this dichotomy to be scrutinised. Indeed, some women do fish on boats (Kaplan 1988; Siason et al 2002; Munk-Madsen 2000; Stella 1996, 1998; Thompson, Wailey and Lummis 1983), although they may not overtly claim "fisherman" status (Yodanis 2000). Some women enjoy the freedom, challenge, lifestyle and potential for making money offered by fishing, and report close relationships when they crew for partners (Kaplan 1988). In other cases, women believe they are capable of the physical labour but choose not to participate in the masculine environment (Howell 2002). Applying a Western public/ private dichotomy in non Western fisheries is also inappropriate as women's domestic and fishery work in these non Western fisheries may be conducted communally in "public" space (Ram 1991; Rubinoff 1999). Land/ sea dichotomies also tend to relate to wild catch fisheries, thus ignoring that much fishing activity occurs inshore or inland, and that women are extensively involved in aquaculture and shore based subsistence fishing (Lambeth et al 2002; Siason et al 2002).

Family business decision making–women in fishing and farming

While a study in the Netherlands reveals that 60% of women are involved in making decisions within their family fishing businesses on issues such as labour, finance and investments (Rana and Choo 2002), there is a lack of international and Australian literature on the involvement of women in decision making in their family businesses. However, we know more about the involvement of women in decision making in their family farming businesses in Australia. Women in farming are less involved than men in farm business decision making (Alston 1995b; Gibbs 1989; James 1990; Rickson 1997) and women's involvement is influenced by the farm being both a business and family institution, tradition and the ideal of generational succession (Gray 1991), women's level of education and perception of marital status (Gibbs 1989), women's relationship to farm ownership (Alston 1995b) and women's identity (Bryant 1999).

Again, in family business decision making, the sexual division of labour may be reflected in the areas in which women participate in decision making in their businesses. For example, Rickson (1997) concludes that men in farming make decisions relating to outside-such as production, machinery purchases, types of diversification-while women make decisions relating to inside, including the social reproduction of the family. There are also penalties for women who breach these traditional roles. Teather (1998) discusses the work of Haslam-McKenzie (1998) who finds that just under half of the women involved in her Western Australian farm survey consider themselves as contributing equally to decision making on their farms. As a result they are feared by the older generation. Teather (1998) comments

> "Such women suffer from the 'daughter-in-law syndrome' (should it be the 'daughter-*out*-law'?) whereby they are excluded from the succession plan or even knowing what provision is made for them in that plan" (213).

The sexual division of labour also operates to position women in fishing who participate in other industry decision making forums. For example, despite the increasing foray by Norwegian women into decision making positions on organising committees for local fishing festivals, these women tend to take responsibility for food preparation and table decorations for these celebrations (Gerrard 2000).

Women in Industry Decision Making and Leadership

Industry decision making–women in fishing and farming

Historically, women in fishing communities have not engaged in formal political processes (Armstrong 1978). While some international research suggests that the participation of women in fishing in industry decision making is gradually increasing (Gerrard 2000; Lambeth et al 2002), in many fisheries and fishing communities the patriarchal nature of fishing continues to restrict women's participation (Allison, Jacobs and Porter 1989; Rana and Choo 2002; Shelly and D'Costa 2002; Siason et al 2002; Smith 1988). A small study commissioned by WINSC (Smith and Sullivan 2001) also finds that there are significant barriers to women's participation in decision making in the Australian commercial fishing industry.

Similarly, women in farming in Australia, and particularly women from diverse cultural backgrounds, are underrepresented in both non-government and government industry leadership positions (Alston 1998e, 2000; James 1989; O'Hagan, Alston and Spriggs 2003; RIRDC/ DPIE 1998; SCARM 2001; Wilkinson 1998a; Wilkinson and Alston 1999). Alston (1998e) surveyed nine industry organisations and found women held 5% of both executive and delegate positions; *Missed Opportunities: Harnessing the Power of Women in Australian Agriculture* (1998) reports that less than 20% of agricultural decision makers are women; at the time of this research only 5% of decision making committee members on the Standing Committee on Agriculture and Resource Management (SCARM) were women, with slightly higher percentages of women on lower level technical and advisory committees (SCARM 2001); and in 1998, the National Farmers Federation Council (NFFC) had one female delegate of a total 41 members (Alston and Wilkinson 1998).

Barriers to women's greater industry participation

The small amount of Australian research on women in fishing has identified barriers to their greater industry participation, including their participation in industry decision making. These barriers include a lack of recognition for their contributions; a lack of time, money and training; conditions on boats; prejudice and discrimination including sexual harassment (Aslin, Webb and Fisher 2000); a lack of childcare; the timing and culture of meetings; inappropriate government consultation; and training costs (McKay, Dimopoulous and Baker n.d.). McKay, Dimopoulous and Baker (n.d.) also find that women repeatedly question the value of their contributions which are not recognised publicly, and are reluctant to participate in meetings where their non-license holding status means they cannot vote. Smith and Sullivan (2001) finds that the barriers to industry decision making for women in fishing include the seasonal and isolated nature of the work; male dominance; industry culture and attitudes;

struggles over power; juggling work and family commitments; a lack of confidence, support and encouragement from within the industry and family; a lack of knowledge of committee procedures; a lack of financial reimbursement for expenses; and board selection criteria based on male networks.

In Australia, research exploring the barriers to the greater participation of women in farming in industry decision making is more extensive (Alston 1998c, 1998e, 2000; Pini 2002a, 2003b, 2004, 2005b; RIRDC/ DPIE 1998). Alston (2000) provides a comprehensive summary of these barriers:

> "…gender relations and the stereotyping of women; a discourse which devalues women and casts them as inadequate leaders; inequitable power relations; the pervasive cultural hegemony; the lack of acknowledgment of women's contribution to the agricultural economy; organisational cultures and practices in farming organisations and departments of agriculture which filter women out; action and inactions of the state; the role of the media; the limited access women have to property ownership because of the practice of patrilineal inheritance and women's point of entry to farming. Of equal importance is the precarious nature of agricultural production and resultant farm incomes which necessitate the sourcing of off-farm income to allow for economic survival. Because women are responsible for this in 80% of cases, their ability to seek leadership is circumscribed by financial stress. Further…government policies have facilitated a withdrawal of health and welfare services from the bush adding to women's traditional caring role and to the pressures on women's time" (Alston 2000, 219).

More specific barriers which have been identified include rural conservatism and masculine culture resulting in conflict with men, sex role stereotypes and the undervaluing of women's work; lack of support from other women and family; family responsibilities including caring for dependents; lack of flexibility in family unfriendly workplaces; lack of role models and mentoring; age and/or physical disability; distance; access, appropriateness and cost of education and training; and women's lack of confidence (Claridge 1998; James 1989; O'Hagan, Alston and Spriggs 2003; RIRDC/ DPIE 1998; Shortall 2000, 2003).

Women in farming are also actively excluded through organisational processes and by those in power who fail to address these inequalities (Alston 1998e; Pini 2002a; Shortall 2001; 2003). Practices, ideologies and power in entrenched organisational cultures restrict women's access to power (Alston 1998e, 2000; Pini 2002a) and reinforce existing gender divisions (Shortall 2001). Women are perceived as lacking the necessary skills to become potential leaders (Alston 1998e, 2000; Pini 2002a, 2003b), and when they do enter decision making, may be perceived as "outsiders" (Pini 2005b). Formal and informal selection processes which further exclude women include: the importance of family background, tradition, and popularity; inadequate board vacancy advertising and campaigning; longevity in office; by-passing the election process; and appointment by "merit" (Alston 1998e, 2000; Pini 2002a). Pini (2002a) finds that the discourse of merit is used to justify women's absence from leadership in the Australian peak sugarcane industry organisation and argues that

> "...it is crucial that the benign and gender neutral status of terms such as 'electoral systems', 'democratic processes' and 'merit selection' be challenged" (75).

In addition, meeting times, sexual harassment, and board structure and function are also reported as barriers to industry leadership positions for women in farming (Alston 1998c; RIRDC/DPIE 1998; Wilkinson and Alston 1999).

Women in leadership positions are seen as "token women" who reinforce the supposed success of existing practices, which positions the responsibility for the lack of women in industry decision making on women themselves, absolving industry organisations of any responsibility to address their gendered practices (Alston 1998e). Alston (1998e) demonstrates this point by referring to a speech made in 1997 by Don McGauchie, the then President of the NFFC who espouses the importance of women entering agricultural leadership:

> "...many rural women are saying 'it's just not worth it', that families, a reasonable lifestyle and good health are more important than a career in corporate agripolitics or the government sector. I suspect they may view the stresses of some of those jobs in the political arena as just too high a price to pay...I would like ... to try and convince you that it is worthwhile, that it is worth it, that indeed you will need to make some compromises that are required to get there" (McGauchie 1997 cited in Alston 1998e, 199).

Similar barriers exist for Culturally and Linguistically Diverse (CALD) and Aboriginal women (Alston 2000; Wilkinson 1998a; Wilkinson and Alston 1999). However, women with diverse ethnic backgrounds are further marginalised in agricultural decision making given their increased invisibility and geographic and cultural isolation; lack of access to transport and communications; language barriers; lack of knowledge about how institutions operate; family censorship limiting women to their traditional roles to a greater extent than Anglo Saxon women; sexism from within women's communities; and sexism and racism from outside women's communities (Alston 2000; Wilkinson 1998a; Wilkinson and Alston 1999). These additional barriers for CALD women are particularly relevant given the ethnic diversity in family fishing businesses in Australia.

Strategies to enhance women's greater industry participation

Australian research identifies strategies for increasing the participation of women in fishing in industry decision making. These strategies include tailoring courses to suit women's needs incorporating flexibility in content and delivery; specific training for leadership, communication and lobbying for industry women; and organisational changes to increase their relevance for women (McKay, Dimopoulous and Baker n.d.). While Aslin, Webb and Fisher (2000) find that many women express interest in women's industry networks to liaise with other industry women and to access training, McKay, Dimopoulous and Baker (n.d.) find that women's organisations are not seen as valuable by the majority of women in fishing in their study, with many women instead indicating a preference for greater involvement in mainstream industry organisations. This difference may be attributable to the diversity between the two studies. The study by Aslin, Fisher and Webb (2000) is a national survey which includes a high proportion of women involved in industry organisations and relevant government agencies, whereas the Victorian state based study by McKay, Dimopoulous and Baker (n.d.) involves focus groups with women in family fishing businesses. Recommendations made by McKay, Dimopoulous and Baker (n.d.) include the targeting of women for involvement in policy development and decision making processes, the development of appropriate training programs and the review of organisational structures to facilitate women's increased participation.

Smith and Sullivan (2001) also identify strategies to increase the participation of women in fishing in industry decision making, including access to personal development and support, committee procedure training and the industry wide implementation of a Charter of Best Practice (Smith and Sullivan 2001). WINSC has actively worked towards implementing these strategies, and has developed and disseminated an *Information*

Booklet in Meeting Roles and Procedures for Regional WIN (Gill 1999) and the *Charter of Best Practice: Seafood Women in Decision Making* (Smith and Sullivan 2001; WINSC 2001).

International research also suggests strategies to counter barriers that women in fishing face to their increased industry participation. These include: providing training which targets women and is designed to accommodate their needs (Lambeth et al 2002; Siason et al 2002); facilitating access to credit and ownership of capital and fishing rights (Williams 2002); developing active grassroots networks for women in fisheries (Nandeesha and Tech 2002; Williams 2002); continued research on women's roles and contributions; the evolution of organisational culture and policy to incorporate gender; development of gender programs in fisheries sectors; increasing women's participation in resource management; and raising awareness that the essential contributions made by women in fishing broaden in times of crisis (Williams 2002).

Drawing again on research with women in farming in Australia, despite the barriers to their greater industry participation, these women are interested in becoming more involved in industry decision making (Alston 2000; Claridge 1998; Haslam-McKenzie 2003b; Wilkinson and Alston 1999). Strategies to facilitate the greater industry participation of women in farming include: leadership awards; leadership courses and training which adopt principles of participation, flexibility, responsiveness and collaboration (Claridge 1998; Haslam-McKenzie 2003b; Muir 2003; SCARM 2001); investigating the potential for enhancing women's participation through information technology (O'Hagan, Alston and Spriggs 2003); educational resources ; and implementing flatter and less hierarchical organisational structures (Alston 2000). Industry organisations can also amend their practices including implementing broader voting criteria, reviewing selection processes, recognising communication and problem solving skills, replacing interviews with a range of alternate processes, and representing a diversity of views on selection panels (RIRDC/ DPIE 1998). Wilkinson (1998) also identifies strategies for increasing participation in decision making by CALD women, which include the establishment of an CALD network to provide personal and professional support including role models, practical and psychological family support, and access to education. These strategies are similar to those identified to increase the participation of women in fishing in decision making, and again, strategies identified by Wilkinson (1998) are particularly relevant given the prevalence of ethnic families in the Australian commercial fishing industry.

Property and economic power

A central theme to emerge from the literature on the barriers to and strategies for increasing women's participation in industry decision making relates to property ownership and economic power. More specifically, women are marginalised in industry decision making by the practice of patrilineal inheritance and individualistic notions of property ownership in both farming (Shortall 1999) and fishing (Aslin, Webb and Fisher 2000; Nadel-Klein 1988; Thompson 1985; Thompson, Wailey and Lummis 1983; Williams 2002).

In farming, patrilineal inheritance limits women's entry to the industry primarily to marriage, means women are unable to independently access finance, and excludes women from a stake in financial ownership, business decision making processes and voting in industry organisations where membership correlates to ownership (Alston 1994a, 1995b; Shortall 2003). Women's financial dependence on their partner exacerbates their marginalised and secondary status (Alston 1994a, 1995b; James 1989; Jennings and Stehlik 2000). Furthermore, these practices also have implications for women who divorce. Shortall (1994) discusses a Canadian case where a woman received a maintenance sum below the poverty line after divorce proceedings, despite 20 years of labour and financial contributions to her farming business.

Similarly, patrilineal inheritance practices exclude women in fishing from inheriting the necessary capital required for fishing (Aslin, Webb and Fisher 2000; Nadel-Klein 1988; Thompson 1985; Thompson, Wailey and Lummis 1983; Williams 2002). While there are some exceptions, such as in North West Spain where women not only fish but also inherit land and boats (Thompson, Wailey and Lummis 1983), men typically inherit the means of production relating to the sea (Nadel-Klein 1988; Ram 1991). Many women are financially dependent on their male partners (Thompson 1985), marginalised in accessing credit (Shelly and D'Costa 2002) and in the ownership of fishing rights and resources (Munk-Madsen 1998a). Women are also excluded from decision making processes where membership of industry organisations is dependent on the ownership of fishing capital (Siason et al 2002).

In Australia, patrilineal inheritance practices perpetuate the occupation of fishing as masculine, and similarly, reproduce farming as a male activity with expectations that sons will continue farming and farms will not be divided (Voyce 1994). While observing that recent legal precedents imply that future inheritance cases could be more sympathetic to daughters, Voyce (1994) suggests that "...as in the past, farmers will work around the reforming impulse of such judgements to the extent the law continues to rub against patriarchal notions of land ownership and inheritance" (80). While McAllister and Geno (2004) suggest that there is a decline in transferring family farm businesses to individuals and instead observe a trend towards family businesses being transferred as collective structures such as trusts and companies, in a similar vein to Voyce (1994), these authors questions whether this will result in more equitable gender relations in farming.

While patrilineal inheritance confers economic power to men, Davis and Nadel-Klein (1992) assert that the Western tendency to equate power with only economic factors potentially overlooks alternate sources of power such as women's leadership and participation in family and fishing and community activities (Davis 1983, 1986, 2000; Gerrard 2000; Nadel-Klein 2000; Skaptadottir 2000). The essential nature of women's labour for ensuring the survival of family fishing businesses, and the necessity for women to engage in decision making when men are at sea, provides the potential for women to hold considerable economic and private power, depending on specific historical, social and cultural contexts (Nadel-Klein 2000; Neis 2000; Thompson 1985). This is the case in several North Atlantic fishing communities (Davis 2000; Nadel-Klein 2000). In Norway, "...women were the 'chief' on shore and men were the 'captain' at sea" (Gronbech 2000, 358), and in Scotland women's activities as fish sellers provides a basis for their economic power (Thompson, Wailey and Lummis 1983). Women also have considerable economic and domestic power in many non Western fishing communities (Ram 1991; Rubinoff 1999; Siason et al 2002; Thompson, Wailey and Lummis 1983; Walker 2001, 2002; Yater 1982).

The Quest for Sustainability – Industries in Context

Over the centuries, fishing and farming have fulfilled a vital role in sustaining populations around the world. However, "the tradition of fishing has been transformed over several decades of human civilization to become a resource extraction industry spanning the entire globe"

(Williams 2002, 147). In Australia, wild catch fisheries are operating under increasingly stringent government regulations in an endeavour to ensure sustainable stock levels (Caton and McLoughlin, 2000; Munk-Madson 2000) and aquaculture is emerging as a major growth sector (FRDC 2000, 2001; NAC 2001). In Australia in the early 1900s, fishing was a small cottage industry conducted by family businesses (Wilkinson 2004). However, as in farming with the advent of agri-business (Bourke and Lockie 2001; Lees 1997), fisheries are becoming increasingly corporatised and the numbers of family fishing businesses are decreasing (Aslin, Webb and Fisher 2000; Wilkinson 2004).

Given the post structural theoretical premise that social relations are located within specific historical contexts, it is necessary to analyse the contributions of women in fishing and their industry positioning within the broader context of an industry in "crisis". Writing on the "crisis" in farming, Rickson (1997) also notes the need for analysis within this broader context to reposition the responsibility for "crisis" to the structural level.

> "...there is a wealth of data and resultant policy focusing upon the individual or family as the unit experiencing, coping, not coping, with the 'crisis'. The rural crisis becomes an individual crisis...that requires ... differing strategies ...from those that would be proposed if the crisis were posited as structural" (Rickson 1997, 116).

This broader focus is particularly pertinent as it is argued that government policy and management agencies are perpetuating the "crisis" in both the agriculture (Jones, Seville and Meadows 2002) and fishing industry (Crean and Symes 1996; Davis 2000; Skaptadottir 2000). Expanding the analysis in this manner requires that attention is given to the practices and discourse of fisheries management and the quest for sustainable fisheries, which may be viewed as a response to the "crisis" in fishing.

Unsustainable industries in "crisis"

Globally, "crisis" is a term used prolifically to describe the current state of fisheries (Kura et al 2004; UN Earthwatch n.d.) and indeed, fisheries management (Crean and Symes 1996; Symes 1996). In Australia, this term is also used to describe the state of both the agricultural industry (Bourke and Lockie 2001) and the fishing industry (*Fishing crisis* 2006). The *Women's Studies International Forum* (2000) edition, which explores how women have coped with the collapse of cod stocks in the North Atlantic fishery crisis argues there is a

> "…need to develop, expand and refine a critical feminist perspective that encompasses the global crisis in the world's fisheries and identifies resources and avenues for gender informed, feminist political action" (Davis and Gerrard 2000, 285).

A major historical factor affecting the social relations of production in fishing was new technology associated with industrialisation. This has transformed the processing sector from being predominately an unpaid, household fishing activity of women, to a factory, wage based processing sector comprising mainly of casualised and low paid women (Nadel-Klein 2000; Porter 1983; Thompson, Wailey and Lummis 1983). In Newfoundland, freezing technology employed in fish processing plants in the 1960s replaced the dried-salt cod industry where women processed family catches (Davis 1986; Porter 1983) and a similar process occurred in England and Scotland in the herring fisheries (Thompson, Wailey and Lummis 1983). These changes have flow on effects: women are no longer visible in unpaid fishery work in Scotland that occurred outside the home (Nadel-Klein 2000); less customary subsistence activities for women in Newfoundland have resulted in a loss of identity (Porter 1983, 1991); and the subsistence based fisher/ farmer lifestyle in Norway is no longer economically viable (Gronbech 2000). Nadel-Klein (2000) argues that the combination of technological and environmental factors that have led to declining fish stocks and greater fishery regulations differentiates the current "crisis" from previous industry insecurity. While individual fishers and their families were previously exposed to risk, currently at stake for fishers is their collective way of life and the increasing probability that their children will "…be forced to 'get out to get on' " (Nadel-Klein 2000, 363).

Similarly in farming, the "crisis" is multifaceted and incorporates economic, social and environmental components intersected by government policy. Australian farming since the 1980s has been characterised by industry deregulation and price destabilisation. Economies of scale and technological practices aimed at increasing efficiency have resulted in the growth of agri-business and the decline of the family farm business (Bourke and Lockie 2001). This has occurred in conjunction with changing household demographics and cutbacks to rural services in a neo liberalist state culture (Rickson 1997). These economic and social imperatives intersect with environmental issues including land degradation, salinity and drought (Alston 2006; Alston and Kent 2004, 2006; James 1989; Lockie 2001b).

The construct of sustainability may be conceptualised as a response to global degradation, including the "crisis" in fishing and farming. As such, it is particularly relevant to this research. The multiple meanings of "sustainability" and "sustainable development" (Achterhuis 1994; Fletcher 2006; Jacobs 1999) also suggest the need to conceptualise and critically analysis sustainability as a contested and discursive construct. The catalyst for international action for the quest to achieve greater sustainability in all areas of life was the release in 1987 of *Our Common Future,* the Brundtland Report (WCED 1990). This led to the development of direction setting documents such as the Rio Declaration and Agenda 21 at the 1992 United Nations Conference on Environment and Development (EA 2007). The term Ecologically Sustainable Development (ESD) was adopted in Australia to ensure that there was a balanced approach in dealing with environmental, social and economic issues and The National Strategy on ESD (Ecologically Sustainable Development Steering Committee 1992) was agreed to by all Australian governments.

Fisheries Policy and Management in Australia

The development of fisheries management policy in Australia has been influenced by economic, social and environmental factors. The following excerpts from the *National Competition Policy Review of Commonwealth Fisheries Legislation* (NCP 2002) provide a succinct overview of the main developments and trends which have occurred in the evolution of fisheries management in Australia.

> "Under 19th century law, territorial boundaries of nations could extend three nautical miles. This applied to Australia and was the extent of jurisdiction of Australian States following Federation. Apart from the South East Trawl Fishery (SETF)-which commenced in 1915-most Australian commercial fisheries were, until the end of the Second World War, localised inshore operations, using small boats and unsophisticated fishing equipment and selling fresh fish to local markets. There was for many years a natural progression in fishing effort that limited the need for management. Apart from laws relating to pearling and whaling, the Commonwealth had passed no laws relating to fishing until the Fisheries Act 1952" (NCP 2002, 9).

> "Following the publication of the then Government's 1989 policy statement New Directions for the Management of Fisheries in the 1990s

the Fisheries Act 1952 was repealed and replaced by the current Fisheries Act 1991 and the Fisheries Administration Act 1991. The focus of these Acts is on fisheries management, with ecologically sustainable development and economic efficiency being the main objectives....Up until at least the mid-1970s the main emphasis of the Commonwealth's involvement in fisheries was directed principally at the development of the industry rather than at management. The same situation applied in most of the States...The initial focus of the Fisheries Act 1952 had been to provide a framework for the development of Australia's fisheries resources. This was modified over the years as management of fisheries gained importance. For example, amendments in 1983 allowed for the creation of limited entry fisheries and, in 1984, allowed for the creation of fishery-specific plans of management....In contrast to the 1952 legislation, the 1991 legislation was directed at managing fisheries..." (NCP 2002, 11).

While historically, fisheries management has primarily been concerned with stock levels, ESD is now a fundamental principle of fisheries management philosophy, policy and practice in Australia (CoA 1991; Fletcher 2002; Fletcher et al 2002; Fletcher et al 2005).

"...ESD is now accepted as the foundation for natural resource management in Australia, being a significant component of fisheries legislation at the state and national level. This represents a fundamental shift in public policy because it affects all government departments and agencies to some degree" (Fletcher 2002, 7)

In ESD, it is the integration of the social economic and environmental factors (often referred to as the "triple bottom line") in decision making which is "...the cornerstone, and the major innovation of ESD" (Fletcher 2002, 15). However, the social component of ESD has not historically been incorporated in fisheries management (Schirmer 2005). The *National ESD Framework for Australian Fisheries: The 'How To' guide for wild capture fisheries* (Fletcher et al 2002), which provides fisheries with a guide to completing ESD reports for use in the national reporting framework, acknowledges the social aspects of ESD as underdeveloped, does not mention gender or women's unpaid work as a feature of industry employment. Similarly, the 1989 Government policy statement *New Directions for Commonwealth Fisheries Management in the 1990s* which led to the *Fisheries Management Act 1991* and the *Fisheries Administration Act 1991* does not address the social impacts of restructuring fisheries or recognise women in fishing as stakeholders. This is despite an explicit commitment to redressing gender inequities in the National ESD Strategy (Ecologically Sustainable Development Steering Committee 1992).

Nonetheless, in more recent publications by the BRS, and due to the involvement of WINSC in the development of these documents, women in fishing are gaining some rhetorical recognition. For example, in *A Social Assessment Handbook for Use by Australian Fisheries Managers in ESD Assessment and Monitoring* (Schirmer 2005), the *Social Assessment Handbook: A Guide to Methods and Approaches for Assessing the Social Sustainability of Fisheries in Australia* (Schirmer and Casey 2005), and the two case studies conducted to inform these documents (Schirmer and Pickworth 2005a, 2005b), women in fishing are either explicitly acknowledged, or there is an acknowledgement that

> "…many fishing businesses include both a fisher operating on a boat and a fishing business partner (paid or unpaid) who manages the financial and administrative aspects of a fishing business" (Schirmer and Casey 2005, 15).

However, gendered practices are evident in the primary focus on men in paid employment in fishing; the primary method suggested for identifying commercial fishers-through license lists and other data held by fisheries management agencies and fish receivers such as cooperatives or processors lists; and the lack of acknowledgement that some women in fishing may not recognise their own contributions or identify as being part of the industry (Schirmer and Casey 2005). Furthermore, in the case study research which informed the development of these guides (Schirmer and Pickworth 2005a, 2005b), few women in fishing participated and focus groups were held primarily in local pubs, clubs and taverns. Indeed, some social impact assessments explicitly perpetuate women in fishing as non stakeholders. This is evident in a social impact assessment of the Western Australian Rock Lobster fishery which refers to women in fishing as "fishermen's housewives" (Huddleston and Drew 2003, 86). The assertion by international feminist authors that women are marginalised in fisheries management policy (Binkley 2000; Gerrard 2000; Munk-Madsen 1998a, 1998b) appears to apply to women in fishing in Australia.

Privatising the Rights to Resources

The current approach to fisheries management can be described as bio-economic and based on Hardin's (1968) "tragedy of the commons" thesis. Referring to wild catch fishing where fish are harvested as opposed to farmed, this thesis argues that unregulated fishers and common property ownership of resources eventually leads to exploitation, as fishers will maximise their fishing activities to secure the most economic gain regardless of potential environmental costs. The thesis also argues that privatisation through the allocation of resources provides the only solution to avoid exploitation. This most commonly occurs through the allocation of Total Allowable Catch (TAC) and Individual Transferable Quotas (ITQs) where fishers are allocated a quota which is a percentage of the overall TAC. Allocating rights to resources is a management "output control", which directly limits the amount of fish that may be caught. In contrast, "input controls" limit fishing effort and include restrictions such as the number of days at sea, boat size and gear type and historically have been the major strategy employed by fisheries management. The shift in fisheries management strategies from input controls to output controls such as ITQs and a TAC is occurring globally due to the failure of input controls to ensure sustainable stock levels (Grafton et al 2005; Grafton, Kompas and Barclay 2006; Jennings, Kaiser and Reynolds 2001; Kura et al 2004). In Australia, the process of privatising the rights to resources has recently occurred in all Commonwealth managed fisheries (AFFA 2003).

Fundamental to the manner in which the rights to resources (in this context, fish) are allocated is the concept of production. Production in fisheries relates to harvested fish tonnage, which equates directly to economic remuneration. The extensive activities of women in family fishing businesses in maintaining fisheries, families and communities are not valued or recognised as productive (Munk-Madsen 1998a). In Iceland, women, families and communities have been absent in discourse on the introduction of the ITQ system (Skaptadottir 2000). In Norway, women who fish are less likely to own boats, more likely to own smaller boats, are categorised as part time fishers, and fish as part of a flexible adaptation strategy where they combine fishing with other spheres of life including family responsibilities and community work (Munk-Madsen 1998a). This pattern of fishing is not recognised as valid professional fishing and excludes women from being eligible for vessel quota (Munk-Madsen 1998a).

> "…modernization and the strengthening of a science based fisheries regulation is a formalization of previously informal patriarchal relations. A common property system with an informal male dominance is being turned into a male property system in gender neutral disguise. Male control is formalized and made less visible at the same time" (Munk-Madsen 1998a, 230).

In Iceland, the introduction of quota has resulted in fewer, larger companies dominating the fishery, the loss of local control over the resource, and local fishers who rent quota are providing profits for owners (Skaptadottir 2000). Furthermore, this restructuring of the fisheries has direct implications for fish stocks (Davis 2000; Skaptadottir 2000).

> "It is now clear that government policy and attempts to streamline the Newfoundland fisheries to make them more commercially productive played an instrumental role in the depletion of cod stocks" (Davis 2000, 352).

Similarly in Australia, within the current industry climate, the structure of the industry is changing. Commercial fishing is becoming increasingly corporatised, the number of owner operator family businesses is decreasing and aquaculture is emerging as a major growth sector (FRDC 2000; FRDC 2001; NAC 2001). An assessment of the social impacts of the introduction of ITQs in the South East fishery finds the fishery has become more economically efficient and there is an "increased race to fish" (Aslin, Connor and Fisher 2001, 4). Furthermore, these authors conclude that the initial hostility from fishers over quota allocation is justified due to flaws in this process, and it remains unclear whether many of the positive expected outcomes have been realised (Aslin, Connor and Webb 2001). Other issues identified as a result of the implementation of quota include reduced numbers of fishing vessels, increased difficultly in entering the fishery, a cultural change from smaller based fisheries to larger corporatised fisheries, small owner-operators being disadvantaged by the increasing concentration of quota, and the corporatisation of fisheries where increasingly fishers are skippering or crewing for quota owners (Aslin, Connor and Fisher 2001). Similarly, a social impact assessment of proposed management in the Western Australian Rock Lobster fishery at the time of this research suggested that these changes would result in similar negative social impacts for fishing family fishing (Huddleston and Drew 2003).

Aboriginal people are further marginalised in the process of the allocation of the rights to resources. While particular commercial fisheries in Australia are exclusively operated by Aboriginal people with special licensing provisions, and joint aquaculture ventures are being established, the increasing commercialisation of fishing is impacting negatively on Aboriginal people through restrictions on fish capture, the exclusion from particular geographical areas due to aquaculture ventures and the increasing costs required to enter commercial fisheries (FRDC 2000). Tsamenyi and Mfodwo (2001) discuss participation-referring to the ownership and access to resources-by indigenous people in commercial fishing. These authors argue that the current positioning of Aboriginal people as traditional fishers needs to be supplemented with discourse on indigenous ownership of commercial fishing rights; the current legislative framework does not support indigenous ownership of commercial fishing rights; and that compensation should be available for indigenous fishers who are excluded from quota allocations (Tsamenyi and Mfodwo 2001). In this context, it can be argued that currently, of more significance than gender equity issues are access issues to commercial fishing for all Aboriginal people.

Research on the allocation of the rights to resources in fishing concludes that this process which centralises the rights according to capital ownership marginalises women (Munk-Madsen 1998a, 1998b) and ethnic groups and crews (Davis and Bailey 1996), and is replicated in Australia (Aslin, Webb and Fisher 2001; Tsamenyi and Mfodwo 2001). The contributions of women in fishing are not recognised in masculine dominated fisheries management policy and are

> "...dismissed in the logic of fisheries management. Gender neutrality in management seems to mean ignorance of the position of women and their diverse and flexible work" (Munk-Madsen 1998a, 231).

Economics, efficiency and inequity

In Hardin's (1968) "tragedy of the commons" thesis, the initial rationale for avoiding exploitation was to ensure sufficient resources to support a growing world population. However, this particular discourse has become silent (Munk-Madsen 1998a). Instead, the privatisation of the rights to resources is acknowledged and positively promoted as a strategy to increase economic "efficiency" (Hannesson 2004). Skaptadottir (2000) identifies this when discussing the inequities which resulted from the introduction of ITQs in Iceland.

> "The minister of fishery argues that the industry as a whole is showing greater profits and is successful. However, the investment in fishing vessels has increased and, in spite of cuts in quotas, fishing has increased because of more fishing in international waters. The privitization of the fish stocks, not conservation, has been the most important goal of the ITQ system and economic efficiency its main measurement" (Skaptadottir 2000, 314).

Similarly, in Australia, Waitt and Hartig (2000) discuss the implementation of the ITQ system in the South East Fishery, and question whether the Australian Fisheries Management Agency's (AFMA) objective of ESD can be attained without social adjustment policies. Noting that AFMA has been successful in its aims–

> "Production has become reorganised. Operators are not only fewer, but also embrace professionalism, quality, efficiency, and the operation of the free market" (Waitt and Hartig 2000)

-these authors argue that AFMA prioritises market needs over ecological factors; that practices of corporate fishing businesses favoured in the ITQ system are potentially the least environmentally sensitive; and that in the implementation of ITQs, there are "winners" and "losers"–the "inefficient", the "unprofessional" and the "part-timer". Recent research suggests that the emergence of the "professional fisherman" or "modern fisherman" further marginalises women (Power 2005; Waitt and Hartig 2005). Further evidence for the social impacts of ITQs in Australia is provided by Bradshaw (2004), who argues that the implementation of quota in the Tasmanian Rock Lobster fishery has resulted in fewer operators, and a significant wealth gap between owner-investors and leasees, who previously were employed in their own fishing businesses.

Beder (2000) argues that policies adopted by governments driven by neo liberalist agendas and environmental economists which attempt to achieve greater sustainability by allocating scarce environmental resources through market mechanisms, removes decision making from communities and reinforces and perpetuates inequities.

> "This suggests that either equity is merely part of the rhetoric of sustainable development and is not really a central concern of those governments, or those governments have not understood the equity consequences of policies being promoted by those who have other agendas and priorities. If equity is to be taken seriously then new ways of decision making must be found....Clearly, merely extending market values to incorporate the environment into existing economic systems will not achieve this" (Beder 2000, 241).

Some authors discuss the need to shift the focus away from stock assessment towards a greater incorporation of human and social dimensions into fisheries management (Maurstad 2000, McGoodwin, Neis and Felt 2000). Incorporating greater human and social dimensions into fisheries management highlights decision making processes as a critical component of "sustainability". Indeed, fundamental in much of the rhetoric of sustainability are notions of institutional reform: decentralised, localised, participatory and collaborative decision making involving all stakeholders including women; and "good" governance which incorporates notions of accountability, equity and social justice (BMU, HBF and UNED Forum 2001; European Commission 2000; OECD 1999). In order to enhance management policy and decision making, there has also been a call for more "democratic resource management" in fisheries (Palsson 2000, 37).

Women's inclusion in all levels of decision making and in all decision making processes is also recognised as central to good governance and to achieving greater sustainability (Braidotti et al 1994; United Nations 1995; Harcourt 1994; OECD 1999). This has also been recognised in Australian government policy relating to women in farming (SCARM 1998, 2001). Yet, despite strategies specifically directed at increasing women's representation, women remain poorly represented in all levels of government and in decision making processes (UNFPA n.d.) including natural resource management in Australia (Davidson and Black 2001). Furthermore, as Shortall (2004b) points out, adopting strategies advocated within more democratic decision making processes such as partnerships, does not necessarily guarantee equitable power relations or the inclusion of women as stakeholders.

Scientific and local knowledge

Bureaucratic structures may be described as patriarchal, state institutions that are underpinned by privileged, masculine, scientific knowledge (Bowling and Martin 1985). Fisheries management bureaucracies are also recognised as possessing these features (Holm 1996).

> "Science is based on the professional creation and certification of knowledge which is tied to powerful interest groups, notably the state, corporations and the scientific profession itself. Patriarchy is based on male control of dominant social structures and the exclusion of women from positions of power....Scientific knowledge is masculine in its neglect of women's experience and its adoption of paradigms built on assumptions of competitiveness and hierarchy" (Bowling and Martin 1985, 1).

Despite the fact that all human activities are situated in a social context (Palsson 2000), the dominant bio-economic fisheries management policy approach legitimates masculine, scientific knowledge and "...popular knowledge, no matter how useful, accurate or systematised is not accepted as scientific" (Bowling and Martin 1985, 17).

> "'Science' is conceptually entrenched as something that must be located among professional workers who are mostly men, situated in the public 'side of the public/private' split. For those subscribing to the conventional conception of science, so congenial to patriarchy, the idea of 'science by the people' is virtually a contradiction in terms" (Bowling and Martin 1985, 17).

This has significant implications for the "local" knowledge of fishers. However, fisheries management policy and practice privileges scientific knowledge over the experiential local knowledge of fishers (Palsson 2000), and this lack of recognition of fishers' local knowledge is a critical factor in the conflict between fisheries management agencies and fishers that characterises many fisheries (Roepstorff 2000). Waitt and Hartig (2000) make the following assertions about fisheries management and sustainability when discussing the introduction of the ITQ system in the South East Fishery in Australia.

> "...all fishers are falsely assumed to be 'rapers' and 'pillagers' of the ocean, with no or little environmental sensibility. In generating this destructive, generic and stereotyped representation of fishing practices, it denies an enormous body of local environmental knowledge. Family fishers spoke of sustainable fishing practices before a formal fisheries management authority existed....government environmental guidelines and quotas replaced a family fisher's environmental sensibilities, driven by the desire of a family inheritance and acquired through years of experience. Undoubtedly this valuable resource has to date been under-utilised in official sustainable management practices" (Waitt and Hartig 2000, 110-111).

Yet, research indicates that the experiential, local knowledge of fishers can significantly contribute to the body of fisheries management knowledge (Gendron, Camirand and Archaneault 2000; Maurstad 2000; Maurstad and Sundet 2000; McGoodwin, Neis and Felt 2000; Palsson 2000; Roepstorff 2000; Rowe and Feltham 2000). Indeed, this has occurred in Canada through the formation of the Fishermen and Scientist Research Society (FSRS) (Zwanenburg, King and Fanning 2000).

The privileging of scientific/expert knowledge also has implications for the experiential knowledge women gain through their daily activities (Bowling and Martin 1985). Indeed, research indicates the potential for women's knowledge to make a vital contribution to natural resource management and sustainable economic growth and development (Sachs 1996). A variety of viewpoints have emerged on the relationship between women and the environment in the ecofeminist literature (see Sachs 1994 for a discussion). Some of these positions are essentialist in their claim that women are privileged to contribute to natural resource management due to women's difference, their greater affinity with nature than men, or their more holistic decision making style. However, other authors locate women's potential to enhance natural resource management as grounded in material relations where women's daily experiences and engagement in specific tasks provide them with partial and situated knowledge (Agarwal 1992; Zweifel 1997). Hassanein (1997) discusses the value of resurrecting the local, experiential knowledge of farmers, including women, which was appropriated by the scientific movement a century ago, and which has the potential to complement scientific knowledge in the quest for more sustainable agricultural industries.

> "Knowledge issues are a critical dimension in the politics of the sustainable agriculture movement….Different experiences in everyday life may create multiple and partial perspectives from which local knowledge for sustainability is generated and exchanged" (Hassanein 1997, 1).

In fishing, while recognising that fishermen possess the knowledge to contribute to fisheries management policy, Power (2000) also identifies the knowledge of women in the seafood processing industry in Newfoundland as a valuable resource to add to the scientific body of knowledge on fisheries management. However, there is no research which suggests that women in family fishing businesses in Australia have experiential knowledge to contribute to fisheries management.

Implications for Women in the Current Industry Climate

There are a multitude of diverse implications for women in the current industry climate in farming, non commercial fishing (see Huang 2002; Medard and Wilson 1996; Ram 1991; Rubinoff 1999; Siason et al 2002; Walker 2001, 2002) and commercial fishing (Aslin, Connor and Fisher 2001; Binkley 1996, 2000, 2005; Munk-Madsen 1998a; Skaptadottir 2000).

Women in farming

Much of the international and Australian research on women in farming is framed in relation to the rural "crisis" and the socio-economic implications for farming families. Research has noted the increasing incidence of women's increasing work loads, stress, involvement in unpaid on-farm work and paid off-farm work, declining physical and mental health (Alston 1995b; Kubik 2005), and the effects of displacement of families and women (Bryant 1991). Indeed, women's increasing labour contributions are so substantial that Kubik (2005) argues that the labour of women in farming in Canada subsidise food produced by farm families and the farm economy. In Australia, criticism has also been levelled at the government's failure to recognise the dramatic social costs of the crisis in farming for families and women which is evident in policies such as asset testing for the receipt of family allowance and the aged pension (Alston 1994b).

Much of the literature explores the shifting parameters of the sexual division of labour in farming families with changes in economic and cultural factors, and observes the trend of women to become increasingly involved in income generating activities to support farming businesses. These activities include off-farm work (RIRDC/ DPIE 1998); on-farm ventures such as small scale craft and hobby cottage industries (Jennings and Stehlik 2000) and more traditional farm diversification and value adding projects (RIRDC/ DPIE 1998). While women may gain personal fulfilment from these activities, in many cases women undertake these activities as a response to the crisis in farming as a strategy to generate vital income to maintain family and farm survival (Alston 1995b; Beilin 1998; Jennings and Stehlik 2000). As fishing becomes increasingly under pressure, women in fishing are also contributing to the survival of their families and family by engaging in paid employment outside their businesses (Binkley 2000).

As a result of their greater participation in off-farm work, women in farming are experiencing increased stress and workloads as paid work is added onto existing farm tasks and domestic responsibilities (Jennings and Stehlik 2000). Women have less time for community activities such as voluntary work; women's labour is diverted away from farming activities; families have the additional costs of women reskilling, retraining and travel; and women experience conflict between their paid and non paid responsibilities and within their personal relationships with their changing roles (Jennings and Stehlik 2000). Women also experience poor physical and mental health due to increased time pressures, the juggling of tasks, a lower standard of living, less leisure time and greater worry (Alston 1994b; James 1989). In addition, the distance required to travel to work and a lack of childcare are problematic for many women in rural areas (Alston 1994b; Dowling 2006b), and many women work in casual, insecure, part time jobs which often do not allow them to utilise their level of skill and training (Alston 1995b; Dowling 2006b). Women's off-farm work also allows men to continue farming-"in a peculiar sense, it is women's off-farm work that keeps farming male" (Shortall 2003, 58). In a similar way, engaging in paid work outside their businesses may mean that women in fishing may be enabling their partners to remain "fishermen". Furthermore, the issues of women in fishing who live in rural areas may experience similar difficulties in accessing paid employment. As in farming, many fishing families are exiting the industry and, as their rural communities decline, are leaving these communities to seek employment (Binkley 2000; Gerrard 2000).

These changes are occurring within a broader context of a pervasive gendered division of labour due to the ideological support for the "natural" domestic centring of women (Alston 1994b; Dempsey 1994; Little 1994; Poiner 1990; Sachs 1983). Instead of the rural "crisis" prompting an equitable redistribution of tasks, women in farming undertake an increasing number of tasks as domestic and family responsibilities remain primarily women's responsibilities (Alston 1995b; Gibson, Baxter and Kingston 1990; James 1989). While women in farming may be seen to be challenging the dominant patriarchal hegemony by contesting new spaces in pursuing off-farm work in the public sphere, they remain subordinated through their continued and increased domestic role and responsibilities (Jennings and Stehlik 2000). The sexual division of labour is also evident in the reallocation of specific farm tasks. When women in farming replace hired labour and become more involved in farm tasks which have previously been defined as masculine, the reallocation of tasks continues to assign more critical farming activities such as sowing crops to men, and less critical farming activities such as ploughing paddocks to women (Alston 1994b). Similarly in fishing, women are increasing crewing for their partners and replacing hired deckhands to reduce business costs (Munk-Madsen 2000).

Women in fishing

Much of the international feminist research on the gendered implications of crisis for women in fishing has explored women coping with the collapse of cod stocks in the North Atlantic fishery that affected 100,000 people and resulted in the restructuring of the industry with significantly fewer fishers (Davis 2000; Gerrard 2000; Nadel-Klein 2000; Neis 2000; Skaptadottir 2000). In Scottish fisheries Nadel-Klein (2000) asserts that "…it is no longer true that a "fisher laddie needs a fisher lassie" to manage a household…" (363). In Newfoundland and Labrador, women accommodated rather than resisted change, and fisheries closures created community conflict which included the contestation of women's historical spaces, increasing violence, a loss of personal and community identity and the reorganisation of family life (Davis 2000; Neis 2000).

Women in fishing are engaging in activities to ensure the survival of their families, businesses and communities. In Newfoundland, women are increasingly undertaking paid employment outside of their fishing businesses and in Nova Scotia, fishing households developed short term household survival strategies such as making clothes (Binkley 2000). To generate expenditure within their communities through tourism, women are redefining fishing community identity by promoting historical fisher women depictions as cultural symbols in signs, statues and museums in Scotland (Nadel-Klein 2000), by holding local fishing festivals and celebrations in Norway (Gerrard 2000), and by establishing handicraft centres in Iceland (Skaptadottir 2000).

Fishers who remain in the industry are increasingly absent from home in an attempt to access resources in richer fishing grounds, and women are becoming stressed as they undertake paid work outside their family businesses while maintaining their domestic responsibilities (Binkley 2000). A further strategy employed by fishers is the pooling of family labour where women act as crews for their partner (Munk-Madson 2000). In many smaller rural communities, younger and middle aged people are also moving away to seek employment elsewhere (Binkley 2000; Gerrard 2000). Women in families who have exited fisheries are also marginalised through gender discrimination in government policies. In Newfoundland, retraining programs and structural adjustment packages excluded women who provided unpaid labour in their businesses, and failed to recognise women who had been made redundant in secondary industries (Binkley 2000).

In Australia, the assessment of the social impacts of the introduction of ITQs in the South East fishery suggests that fishing families are increasingly under pressure and likely to exit the industry (Aslin, Connor and Fisher 2001). Other research also indicates that women are concerned about the implications for themselves and their children of the capital intensive development in the industry (McKay, Dimopoulous and Baker n.d.; MyriaD Consultants 2000; Stella 1998), with women requiring a higher level of numeracy and literacy skills and greater time to invest in the increased administration associated with the system (Aslin, Connor and Fisher 2001).

Women's industry activism

Historically, women in fishing have demonstrated activism For example, in Scotland in the early 20th century, women fishery workers became unionised, and the "gutting quines" or seasonal herring workers held recurrent strikes for improved wages and conditions from 1913 to 1953 (Thompson, Wailey and Lummis 1983). More recently, women in both fishing (Gerrard 2000; Nadel-Klein 2000; Skaptadottir 2000) and farming (Hogan 1994; Liepins 1998; Meyer and Lobao 1994; O'Hara 1994; Sachs 1994, 1996; Teather 1998) are demonstrating activism in response to the "crises". In the broader context of the women's liberation movement, the Australian women in agriculture movement emerged as a response to the rural crisis of the 1980's. This movement has successfully increased the visibility of women in agriculture and broadened the traditional farming agenda, primarily due to the networking between rural women, corporate women, researchers and links into government bureaucracies (Alston 2003; Ferrari and McKinnon 1998).

The women in agriculture movement includes the formation of organisations such as Australian Women in Agriculture (AWiA) in 1993, which aims to raise the profile of women in agriculture, address inequalities and work to ensure the future of agriculture (Teather 1992). Events have been and continue to be held such as the International Conference on Women in Agriculture in 1994 and annual Women on Farm Gatherings. Government rural women's units were established in agricultural departments (Alston 2003; Haslam-McKenzie 2003b), and a policy agenda developed (AFFA 1999; SCARM 1998). Attention has been focused on issues such as property transfer, women's legal status and formal legal structures which have reinforced patriarchal rural ideologies

(Voyce 1994; 1997). This movement has challenged the patriarchal nature of established farm organisations and has instigated ongoing relations with these groups, with one of the most significant developments occurring when the Victorian Farmers Federation changed their constitution to allow two votes per farm membership instead of one, thereby giving women partners an independent voice.

Similarly, in the past decade, the visibility of women in fishing has increased through the emergence of organisations for women in fisheries around the world (often in response to the "crisis") including America, the Pacific (Lambeth et al 2002) and Asia (Matics, Poeu and Siason 2002; Siason et al 2002). This collective activism by women has also occurred at a more local level. For example, in Norway, women have lobbied government, formed women's groups and criticised public policy (Gerrard 2000). The growing awareness and activism of women in fishing is also reflected in the increasing prevalence of international and national events, workshops and conferences (Nandeesha and Tech 2002; Williams et al 2002). The first global workshop on Women in Aquaculture was held in Rome in 1987 and organised by the Food and Agriculture Organization (FAO) (Nandeesha and Tech 2002). This was followed by the 1998 Chaing Mai and the 2001 Taiwan women in fisheries symposiums organised by the Asian Fisheries Society, and the global workshop on "Gender, globalisation and fisheries" in 2000 facilitated by Barbara Neis (Williams 2002). Clear messages that emerge from these events include: the need to value women's extensive and diverse contributions, particularly in the context of globalisation and crisis; the need for more equitable access to property; the greater participation of women in decision making and further research relating to women in fishing (Williams 2002).

In Australia, WINSC was established in 1999 as a national body for women involved in the fishing industry (WINSC n.d.). This organisation is proactive in promoting women in the fishing industry, holds an annual conference and regular workshops for its members and undertakes action to further the industry generally. WINSC has gained significant prominence in a relatively short time frame and has had representation on the Regional Women's Advisory Council (RWAC) and the Rural Women's Secretariat (RWS).

While there is been no scholarship specifically addressing women's involvement in, resistance or activism through fishing organisations, there is a significant amount of research conducted on rural and farm women's organisations. This research explores: the differing philosophies behind organisations (Teather 1992, 1994a, 1994b,); the motivations of women who become members; the processes by which women become empowered through these organisations (Liepins 1998); and the implications of categorising these organisations in terms of funding and conceptualising the integrated nature of farm women's lives (Shortall 1994).

In Australia, the increased visibility of women in farming on the national government agenda is a significant development (Haslam-McKenzie 2003a). However, the manner in which the government has constructed women represents a shift away from the feminist tenets of justice and equity (Alston 2003). In government rhetoric, increasing women's representation and recognition is not articulated as necessary on social justice grounds, and instead, women are positioned as a resource with a potential which is currently untapped. Pini (1998) asserts that the rhetoric on women in farming in government documents from approximately 1995 reflects this dominant economic rationalist discourse, including in *Missed Opportunities: Harnessing the Power of Women in Australian Agriculture* (RIRDC/ DPIE 1998).

> "Once again, the premise is that women can be utilised to assist in dealing with these hard times. Indeed, women can be 'harnessed' as one might harness a horse or other draught animal. To not harness women as such are 'missed opportunities' – not for women, but for the agricultural sector" (Pini 1998, 227).

This construction positions women in farming as "…the means to an economic end" (Pini 1998, 231). While there are significant benefits of quantifying women's labour, adding women on to the existing narrow economic model fails to reconceptualise "value" in other than monetary terms (Pini 1998). As Waring (1998) comments:

> "…economics has become a great coloniser. When women have said we are unpaid and we need to be visible, they have raced around and said, well we will impute the value of all your unpaid work, and we will superimpose it on this myopic unidimensional system. You can sit there along with your chemical weapons, and other chemical poisons, and feel good about the fact that you now count. As if somehow or other that makes us feel great" (Waring 1998, 38).

While acknowledging that employing this economic rationalist discourse may be a deliberate strategy of femocrats to keep women in farming on the agenda, Pini (1998) argues that the terms of the debate have now been fixed and that there are potential costs. Women will become increasing overloaded as they add more tasks to their schedules and, as the responsibility to become more involved lies with women, structural and organisational change is not addressed (Grace and Lennie 1998). Furthermore, women may become redundant when there is no longer an economic imperative behind the rhetoric (Pini 1998).

Furthermore, under the neo liberalist government in office at the time of this research, there had been a subtle withdrawal of support for women's units in agricultural departments through mainstreaming, name changes and/or cutbacks (Alston 2003; Haslam-McKenzie 2003a). Mainstreaming may be understood as a radical process which critically reviews women's marginalisation and where gender is incorporated across redesigned department structures and processes (Alston 2003). When this conceptualisation of mainstreaming of rural women's units is implemented, it has the capacity to effect real change across institutions and organisations (Alston 2003). However, bureaucracies are hostile to mainstreaming gender (Bhatta 2001) and the potential for change is contingent upon the conceptualisation of mainstreaming that is adopted (Alston 2003). In this context, to avoid women being swept of the agenda entirely, Alston (2003) argues it is imperative to maintain existing women's units. This positioning of women in farming in government rhetoric and processes is pertinent when considering women in fishing as there is a lack of research on these issues, an international call to shift the emphasis from women to gender in fisheries (Chao et al 2002; Williams et al 2002), and significant implications for what women are expected to do. Furthermore, there is the issue of who is responsible for addressing the marginalisation of women in fishing.

Conclusion

This literature review provides evidence that women in farming have been and remain largely invisible within a masculine dominated industry. Women in farming are largely invisible in the media and industry statistics, are underrepresented and actively excluded from industry decision making, and lack stakeholder status. Yet, despite this, women make significant contributions to their industry, businesses, families and communities, often in the contexts of their industry in "crisis". Many of these findings are replicated in research with women in fishing overseas.

However, research with women in the Australian commercial fishing industry is quite recent and fragmented, and many issues require further investigation (Aslin, Webb and Fisher 2000; McKay, Dimopoulous and Baker n.d.; Stella 1998). Further attention needs to be directed towards exploring the extensive contributions of women in fishing in Australia and their status as industry stakeholders. This is particularly important to ensure that women are not further marginalised in change processes occurring in the context of the current fisheries "crisis", and are able to contribute to decision making concerning the future of their industry.

CHAPTER TWO

FEMINIST THEORY

Introduction

In this chapter, I outline the feminist theory that is employed to explore the lack of visibility of women in fishing, their contribution to the Australian commercial fishing industry, the barriers to and opportunities for their contribution and the implications for women in fishing in the current industry climate. I discuss the broad theoretical approach adopted in this book and provide an explanation of the post structural concepts critical to gaining an insight into the contributions of women in fishing. After a brief definition of feminism, I explain the theoretical position adopted in this research, which draws from both feminist informed Foucauldian post structuralism and standpoint theory.

The chosen theoretical framework follows the approach of other feminist writers, who argue for reconciliation of the ontological tension between the constitution of reality from experience (standpoint theory) and from discourse (post structuralism) by locating knowledge as historical, situated and partial (Haraway 1988; Hirschman 2004; Weeks 2004). This approach allows for the recognition of diversity between and amongst women, the exploration of women's subjectivities at both an individual and collective level, and an analysis of macro and micro power relations. Employing this approach also allows both the material conditions which ground women's experience to be considered, as well as the way power/knowledge is constituted in and through women's subjectivities.

This is followed by a discussion on "social positions", where I explain my research focus on the categories of sex/gender and race/ethnicity, and my omission of other social positions including class. I then provide a detailed exploration of my interpretation and application of the key theoretical concepts critical to gaining an understanding of the contributions of women in fishing and the positioning of women in the dominant industry discourse-power, knowledge, discourses, subjectivities and subject positions.

Defining Feminism

Feminism is central in this research and can broadly be defined as "…a critique of misogyny, the assumption of male superiority and centrality" (Beasley 1999, 4). This central tenet of feminism is also often accompanied by a focus on gender, a link between theory and research (Kelly, Burton and Regan 1994), and an historical association between theory and political action (Ramazanoglu 1989 in Poole 1997). Feminism can also incorporate "…an emotional attachment for those who claim membership within it" (Beasley 1999, 118). However, there are diverse theoretical strands or "feminisms" incorporated within the broader school of feminist thought which are informed by differing epistemological and ontological positions (Humm 1995; Poole 1997). This influences which research questions are asked, leads to differing kinds of analysis and knowledge (Campbell and Wasco 2000; Maynard 1994), and determines the use of concepts and language (Poole 1997).

Beasley (1999) suggests that it is the central characteristics and the level and nature of departure from traditional thought that characterises the boundaries in any definition of feminism, and argues there is no need for definitions of feminism to reflect a rigid unity. Indeed, diversity in claims to knowledge production act to validate the sentiment that feminist research is not about replacing the dominant positivist tradition with a singular feminist strategy for eliciting truth (Ardovini-Brooker 2001; Harding 1987a, 1987b).

This book employs the following conceptualisation of feminism offered by Weedon (1989), which is additionally informed by feminism being an "…historically and culturally specific set of practices in social situations where women have conflicting interests and diverse experiences" (Ramazanoglu 1989, 22 in Poole 1997, 96).

> "Feminism….is a politics directed at changing existing power relations between women and men in society. These power relations structure all areas of life, the family, education and welfare, the worlds of work and politics, culture and leisure. They determine who does what and for whom, what we are and what we might become….political questions should be the motivating force behind feminist theory which must always be answerable to the needs of women in our struggle to transform patriarchy. As feminists

we take as our starting point the patriarchal structure of society. The term 'patriarchal' refers to power relations in which women's interests are subordinated to the interests of men....Patriarchal power rests on the social meanings given to biological sexual difference. In patriarchal discourse the nature and social role of women are defined in relation to a norm which is male" (Weedon 1989, 1-2).

Reconciling Feminist Standpoint Theory and Post Structuralism

This section explains the broad theoretical position I have adopted in this book and draws from both feminist standpoint theory and feminist post structuralism. I commence with an overview and critique of each theoretical paradigm and employ the concept of "situated knowledge" (Haraway 1988) as an approach to reconcile the tensions between these views. This is followed by a justification of the central focus within this research on sex/gender and race/ethnicity and my exclusion of additional social positions including class.

Feminist standpoint theory

Standpoint theory emerged during the late 1970's and 1980's as a critical theory, philosophy, methodology, epistemology and political strategy to empower oppressed groups. Focusing on relations of power, and the historical and social locatedness of knowledge (Harding 2004b), it extends feminist horizons by exploring a more accurate knowledge derived from women's social location and experience (Hartsock 1987). Standpoint theory also challenges the dominant masculine scientific discourse on knowledge, objectivity and positivist method which had previously denied the actualities of women's experience and supported hegemonic masculine practice and ideology (Harding 1987b, 2004c; Smith 2004a).

Authors emphasise different resources for knowledge for diverse groups-women's responsibility for daily lives (Smith 2004a), women's subsistence activities under capitalism (Hartsock 2004a) including in the third world (Mies and Shiva 2004), and Black women's lives (Collins 2004a; hooks 2004). Despite these variations, central prepositions of standpoint theory include that: women's experience is shaped by everyday material life, which infers difference between the experiences of men and women; privilege is least visible to those who control the resources that define dominant ideology; it takes the standpoints of the oppressed to recognise systems of oppression and privilege; and that research must begin from women's lives, i.e. "the view from below", as women's location is a resource for the construction of new knowledge (Harding 1987b). The importance of the material and historical conditions which are precursors for social structure and which in turn inform experience are central to feminist standpoint theory. Indeed

> "...women's life activity does form the basis of a specifically feminist materialism, a materialism which can provide a point from which both to critique and to work against phallocentric ideology and institutions" (Hartsock 2004a, 49-50).

A standpoint differs from a perspective or viewpoint. Instead it is a political achievement where oppressed groups have knowledge of their own project, with a standpoint being achieved through struggle with dominant hegemonic masculine practice and discourse (Harding 2004b). The knowledge of oppressed groups (or the view from below) provides both the resource of distinctive experience for oppressed groups and the *potential* for liberation, with standpoint projects assisting in producing "...oppositional and shared consciousness in oppressed groups..." (Harding 2004b, 3). For Collins (2004) writing about Black women, the relationship between oppression, consciousness and action is dialectical. Oppressive structures create patterns of choice, which are perceived in differing ways. Black women's consciousness, defined as "...their analytical, emotional, and ethical perspective of themselves and their place in society" (115), is crucial for their involvement in potential activism, which in turn affects perceptions of structural choices, influences actions and alters the oppressive experience. Collins (2004) also comments that the processes of changing individual subjectivities and a group coming to voice are inherently different, as the former involves conflict amongst the self, while the later involves conflict within the group.

Post modern/ post structural feminism

Emerging in the 1980's, post modern feminism represented a theoretical shift in focus from equality to difference and from universality to plurality. Employing the relativist position that it is not possible to produce one "truth" about women's position in society enabled diversity amongst women to be recognised and, differences between, and within, subjects to extend to more than gender. However, this approach raised the significant challenge for feminism to reconcile relativism with political action (Fawcett and Featherstone 2000). The rejection of "truth" as a foundational reality has lead many post modern feminists to focus on the fluidity of subjectivities, and how meaning is constructed through multiple representations of masculinities and femininities in cultural texts (Beasley 1999; Poole 1997).

While it is difficult to precisely define post modern feminism (Fawcett and Featherstone 2000) and the term is often interchanged with post structuralism, post structuralism refers specifically to the critical theoretical response to structuralism (Beasley 1999). Particular theorists influential in the development of post structural feminism include Foucault, who theorised power as capillary, knowledge and power as unified, and recognised the body as an historical and cultural entity where sexuality is produced to structure social relations. Language is also fundamental to post structuralism. Lacan, Derrida and Lyotard emphasised language as being critical for the development of one's sense of self and subjectivity (Poole 1997), with Saussure proposing that there is no fixed underlying structure to language, and that systematic difference (or differentiation of language) is the precondition to meaning. Lacan's theorising that the subject "I" is formed from a split between the unconscious (associated with loss of the Other/mother) and conscious, and that self is never one coherent unity, destabilises Saussure's notion of a fixed underlying order to language, rendering meaning/language socially contextualised, constructed and never neutral (Beasley 1999). Foucault drew from Saussure's interest in systems of meaning and applied this to social life, asserting that one "truth" enforces dominance, and that power is exercised in all social relations (Beasley 1999). Language, meaning and power are intertwined and constituted in discourse.

> "...poststructuralists usually perceive meaning (conceptions of truth, the forms of power relations) as being neither entirely arbitrary-since particular meanings are socially legitimated while others are marginalised-nor absolute or eternal. Meaning is not random but also not fixed since it is constantly being produced within particular contexts. Poststructuralists tend to stress the shifting, fragmented complexity of meaning (and relatedly of power), rather than a notion of its centralised order" (Beasley 1999, 91).

Much post structural feminist writing employs deconstruction as a technique to challenge binary oppositions such as man/woman, culture/ nature and mind/ body. These binary oppositions are constructed hierarchically within specific cultural contexts and in relation to specific theories. For example, public/ private relates to functionalism, and production/ reproduction reflects the Western distinction between the production of things and people (Moore 1993). This dualistic tendency is a feature of systems of oppression, with either/or dualistic constructs categorising people *in relation to their difference to each other*. Each half of these inherently unstable dichotomies only gain meaning in relation to its difference from the other term. Furthermore, each half of the dichotomy is perceived as not complementary, but as being in opposition to the other half, thus suppressing alternatives (Hughes 2002).

> "Dichotomous oppositional differences invariably imply relationships of superiority and inferiority, hierarchical relationships that mesh with political economies of domination and subordination" (Hughes 2002, 110).

These binaries are also permeated with a *sexual* hierarchy even when appearing to be gender neutral (Beasley 1999).

Deconstruction has resulted in the destabilising of the category "woman" with feminists re-evaluating the assumptions that biological sex is the basis for gender differences, and that biological categories exist prior to culturally determined gender constructs (Moore 1993). The social construction of binary sex categories differs from the physical variations in human populations. Neither differences in physical attributes nor the differing reproductive roles of women and men automatically produce the discrete binary sexed categorisation evident in Western discourse (Moore 1993). As such, this has resulted in a move away from cultural gender representations being based on biological sex differences, to an analysis of how cultures construct differences between women and men. "Woman" is not a stable and fixed identity, and there are differences within and between women (Moore 1993).

> "Multiple forms of difference – race, class, gender, sexuality – intersect within individuals, and identity is therefore premised on difference" (Moore 1993, 204).

Therefore, in contrast to identity politics, the fixity of all categories including sex/gender, race and class are challenged and destabilised. With no inherent category "woman", theorising focuses not on universalising assumptions about women's domination, but on destabilising power relations (Beasley 1999). Politics in this perspective may be defined as:

> "...a local and situated activity undertaken by discursively constituted subjects. Political resistance, furthermore, is defined as challenging the

> hegemonic that writes a particular script for a certain category of subjects. Resistance is effected by employing their discursive formations to oppose that script, not by appealing to universal subjectivity or absolute principles" (Hekman 2004, 234).

Hughes (2002) offers the term "conceptual literacy" in recognition of the diversity and plurality in language and the political contestation and implications of meanings.

> "I am suggesting neither that there are differences of opinion about concepts which possess an uncontestable core, nor that concepts are linked to incommensurable theories. Rather I see concepts and categories as shaped by political goals and intentions. Contests over the meaning of concepts, it follows, are contests over desired political outcomes" (Bacchi 1996, cited by Hughes 2002, 11).

Post structuralism, with its focus on power, subjectivity, language and discourse, is seen as a way of bridging the personal and political and understanding patriarchal relations as more than structural. Indeed, post structuralism is a

> "…theory which can explain how and why people oppress each other, a theory of subjectivity, of conscious and unconscious thoughts and emotions which can account for the relationship between the individual and the social" (Weedon 1989, 3).

A feminist post structural/ standpoint theory conversation

While post structuralism and standpoint theory share a social constructionist position where the nature of the subject is not pre-given but socially constructed, there remain some points of tension between the two positions. These relate to difference, essentialism, women's group consciousness, claims to knowledge, emancipatory goals, and the ontological tension between reality being constituted through experience grounded in material relations (standpoint theory), and through discourse (post structuralism).

Difference, essentialism and women's group consciousness

The essentialist critique of standpoint theorists' construction of men and women as "sex classes" derives from its Marxist origins (Harding 1987b; Hartsock 1987). The earliest standpoint theorists did acknowledge difference amongst women (Hartsock 2004a; Jaggar 2004). However, more recently standpoint theorising has been enhanced by Black/ coloured and Third World feminists, who propose coalition politics based on an understanding of gender oppression that intersects with other social positions such as race, class, religion and sexuality (eg, Collins 2004a; hooks 2004). This perspective has also been adopted by Moreton-Robinson (2002) who argues that Whiteness is the dominant, assumed and ignored feature of Australian feminist writing. There is a need to move beyond the "add race and stir" conceptualisation of difference, and to gain deeper insight into how class, race gender and sexuality are used to construct each other (Collins 2004a).

> "This logic of multiple subjects leads to the recognition that the subject of liberatory feminist knowledge must also be, in an important if controversial sense, the subject of every other liberatory knowledge project. This is true in the collective sense of 'subject of knowledge', because lesbian, poor, and racially marginalized women are all women, and therefore all feminists will have to grasp how gendered, race, class, and sexuality are used to construct each other" (Harding 2004c, 134).

However, standpoint theorists' anti-modernist version of de-centred subjects with no singular "truth" (Harding 2004c) has led to the criticism that it represents a relativist position involving a plurality of unlimited positions (as the experiences of all women differ), implying no possibility for a unified stance (Hekman 2004). Several authors have argued that this is a serious misreading of standpoint theory that fails to recognise its political intent (Collins 2004b; Harding 2004d; Hartsock 2004b) and overlooks the fact that standpoint theorists' emphasis on macro power relations represents a focus on a different aspect of the subject (Hartsock 2004b).

> "...standpoint theory places less emphasis on individual experiences within socially constructed groups than on the social conditions that construct such groups....[It is the] common location within hierarchical power relations that creates groups, not the results of collective decision making within the groups. Race, gender, social class, ethnicity, age and sexuality are not descriptive categories of identity applied to individuals. Instead, these elements of social structure emerge as fundamental devices that foster inequality resulting in groups" (Collins 2004b, 247-8).

Collins (2004) argues that groups, while having fluid boundaries, have a permanency over time and share an historical commonality of experience. However, this does not mean that all individuals share or interpret experiences similarly, and so employing the construct "group" does not subsume all difference. Groups are not reducible to aggregates of individuals.

> "…the constitution of the subject… is the result of a complex interplay of "individuals" and the larger-scale social forces....Moreover, the constitution of the "collective subject" posited by standpoint theories requires an always contingent and fragile (re)construction/transformation of these complex subject positions" (Hartsock 2004b, 245).

Claims to knowledge

Standpoint theory has also had to counter the critique that it privileges certain knowledge. Hekman (2004) argues that the plurality inherent in viewing knowledge as situated does not lend itself to the privileging of certain perspectives over others, and suggests instead that the reality of women's lives is a discursive formation which is partially constructed by standpoint theorists. Standpoint theorists have responded to this criticism by arguing that it fails to recognise the privileging of knowledge as a political strategy based on ethics and justice (Collins 2004b; Harding 2004d; Hartsock 2004b). In addition, other areas such as medicine are based on specific values and interests which subsequently shape research processes and outcomes; knowledge claims only have meaning in specific cultural contexts; and decisions are made on all available yet incomplete evidence (Harding 2004c). Furthermore, it is inadequate to assert that the modernist claim to knowledge is superior if all knowledge-including the modernist discourse-is in fact socially located and influenced by local cultural values (Harding 2004c).

Emancipatory goals

Post structural theorists' adoption of deconstruction as an interrogatory principle lies in contrast to standpoint theorists' aim of bringing women's group consciousness into being. Indeed, a significant concern is the deconstruction of the category "women", which negates generalisations essential for political action. This deconstruction has not only reinforced the notion-incorrectly, according to standpoint theorists-that group and collective consciousness are the same, but has also resulted in a shift in post structural feminist research away from a focus on women's lives towards greater theoretical discussion (Maynard and Purvis 1994). This literature is often characterised by a level of abstraction, which, while acknowledging women's oppression, is not grounded in the daily lives of women or concerned with changing institutional power relations (Collins

2004b, 252). Standpoint theory, like liberal feminism, *is* concerned with modernist emancipatory aims. In contrast to post structuralism, these epistemologies not only aspire to enhancing the objectivity of science (Harding 1987b), but are imbued with an optimistic sense that overcoming oppression is possible (Fawcett and Featherstone 2000).

Reality–experience vs discourse

The ontological tension between post structuralism (where reality is a discursive formation) and standpoint theory (where reality is derived from materialism which informs experience) raises central questions about the nature of experience. Standpoint theorists argue that people have a tacit knowledge about their everyday lives, which is only transformed through telling into knowledge to become part of the language game of experience (Ruddick 2004; Smith 2004b). This tacit knowledge is

> "...a knowing that is the very texture of our daily/nightly living in what we know how to do, how we go about things, and what we can get done. We know how and where to go shopping; we know how to read a book in the less-than-aware dimensions of turning pages from left to right; we know washing dishes, sweeping floors, cleaning; we know putting on makeup and washing our hair; we know how to recognize the boundary between street and sidewalk…we know as a matter of doing. This is a knowing that is of the socially organized ground of our participation in living with others, some of it, indeed, altogether beyond consciousness, but no less what we know how to do…....taking women's standpoint and beginning in experience gives access to a knowledge of what is tacit, known in the doing, and often not yet discursively appropriated (and often seen as uninteresting, unimportant, and routine)" (Smith 2004b, 265-266).

Standpoint theorists argue that experience is more than the language game often reflected in post structural literature (Smith 2004b). Everyday experience is imbued with social relations and standpoint theorists are concerned with the "real" nature of the experience of oppression.

> "While I respect postmodern contributions in deconstructing languages of power, standpoint theory encompasses much more than changing the "language game of politics"….Oppression is not a game, nor is it solely about language-for many of us, it still remains profoundly real" (Collins 2004b, 253).

Overcoming the modern/ post modern binary

Some feminists have reconciled these epistemological and ontological tensions and drawn from both standpoint theory and post structuralism by deconstructing the fixed choice inherent in a rigid modern/post modern binary (Haraway 1988; Hirshmann 2004; Weeks 2004).

> "…the modernist tradition (with its wealth of differences) was equated with a paradigm that perhaps most closely resembled certain themes within Enlightenment modernism, poststructuralism (in all its heterogeneity) was identified with something called postmodernism, and the rest of us were compelled by the relentlessly oppositional logic used to maintain this dualistic framework to choose either one side or the other….This reactive dynamic left us with two equally unsatisfactory choices: a reaffirmation of the humanist subject in some form or the death of the subject tout court, a voluntaristic theory of political agency or a thoroughgoing social determinism. What now remains in the wake of this particular formulation … is a history of distortions and a continuing inability by many to recognize the potential of, and nuances within, the rich traditions of thought that were subsumed into the category of modernism and the equally valuable theoretical frameworks that confined by the category of postmodernism" (Weeks 2004, 181-2).

Weeks (2004) employs standpoint theory to explore women's labouring practices. She argues that women's labour is "…one of many constitutive links between social structures and subjectivities" (184) which as a category can be employed for political purposes. As standpoints are constructed around subjectivities deriving from labouring practice, collective subjectivities are defined by practices and desire, and not knowledge and consciousness. This theorising of subjectivity around labour moves away from the deterministic/voluntaristic dichotomy and recognises labour as both a site of oppression and resistance.

> "This potential power, these alternatives, are located not in some natural or metaphysical essence, but in our practices; more specifically, these possibilities of feminist subjectivity, of feminist political agency, are grounded in an ontology of laboring practices" (Weeks 2004, 188).

Following this practice of drawing from both standpoint and post structural theory by these feminist authors, and other authors who advocate for a "situational" approach in reconciling tensions between structural and discursive approaches (eg, Connell 2003), I employ Haraway's (1988) concept of "situated knowledge". This concept recognises all knowledge as partial and situated, and employs the concept of "vision" (as opposed to "truth"), to overcome the modern/post modern binary, where vision from below is neither a relativist or essentialist view, with both these approaches being "god-tricks" (Haraway 1988). The advantage of this conceptualisation is that a standpoint is not conceptualised as starting from women but as starting from difference, and as such, is able to retain a feminist political integrity.

> "...the alternative to relativism is not totalisation and single vision...The alternative to relativism is partial, locatable, critical knowledges sustained through the possibility of webs of connections called solidarity in politics and shared conversations in epistemology..." (Haraway 2004, 89).

Social Positions-Sex, Race and Class

A term which may be employed to describe common fixed broad ranging socially constructed categories such as sex/ gender, class and race is "social positions".

> "Social positions are based on structural organization such as class, race and gender which circumscribe and access movement into certain subject positions. These structurally organized social positions enable and limit our access to cultural, economic, social and symbolic capital and thus the ability to recognize ourselves as the subject positions we occupy. (Dis) identifications from/with and (dis)simulation of these social and subject positions are the means by which identities come to appear as coherent" (Skeggs 1997, 12-3).

In addition to gender, race and class, there are a multitude of additional socially constructed positions including age, disability and sexuality. Furthermore, in contrast to the "add woman, race or class and stir" approach, these social positions are intertwined and impact on each other, and interact on an individual's subjectivity. People identify themselves to varying degrees as being part of, or not part of, these categories (social positions), depending on each individual and how they are exposed to alternate discourses. The subjectivity of one body cumulatively reflects the adoption and negotiation of, and resistance to, the multiple subject positions associated with these social positions.

In order to have some political impact in a Western world which espouses the construction of these fixed social positions as concrete, it becomes necessary to declare one or more of these social positions as the main area of investigation in order to both "start somewhere" and maximise the political implications of the work. To clarify this further, I imagine that the intertwined, shifting and complex nature of the constituting and constituted nature of individuals' subjectivities in relation to various social locations is only possible by analysing narratives from one or very few individuals.

In this research, I focus on the social positions of sex/ gender (women and men) and race/ ethnicity, and attempt to explore the similarities and dissimilarities of experience within these social positions. However, I acknowledge that in any one body, other social positions (many of which I would be unaware of), simultaneously impact on each of these other locations. As such, while speaking of the commonalties of shared experiences, I am committed to recognising both the inseparable nature of various social positions and the diversity "within" these social positions.

In regard to "sex" (women and men), I understand that signifiers prescribing cultural meanings are "written" on physical bodies and, as with "gender" (femininity/ masculinity), that these binaries are constructed both relationally and hierarchically. In relation to race/ ethnicity, I employ the term ethnicity in order to avoid the biological determination implied in "race". To capture the multiple possibilities within this social position, when attempting to gain an understanding from women in fishing of their ethnic/ racial identity, I asked them to describe their ethnic/ cultural background.

I am also compelled to respond to the charge from Moreton-Robinson (2000) who argues that Whiteness is the dominant, assumed and ignored feature of Australian feminist writing. I believe this is the case and attempted to include Aboriginal people in this research. However, at the time of this research, Aboriginal people were not involved in commercial fishing as owner operators in family businesses. Aboriginal people were absent in the commercial fishing industry discourse, with instead, the use of the term "Aboriginal" prevalent in discourses of traditional fishing. Therefore, the absence of Aboriginal people from this research does not result from my identity as a white, middle class feminist, but instead results from the more systemic exclusion of Aboriginal people from commercial fishing (Tsamenyi and Mfodwo 2001). This absence also highlights the need for further research to explore the exclusion of Aboriginal people, including women, from commercial fishing in Australia.

Skeggs (1997) also argues that class has fallen off the feminist agenda and that post structural theorists have tended to dismiss this social position. While she acknowledges the extensive difficulties in developing practical definitions of class, her charge is that class is a major feature of subjectivities, which reflects a material reality *which exists* and that its exclusion, particularly by white, middle class feminists, silences those who are less privileged.

> "... do we mean class structure, identity, consciousness, action, and so on when we speak of class? Other difficult questions are also raised: how does class relate to the sexual division of labour, and is it a cause or an effect? Have feminists avoided class because it is impossible to measure accurately?....Or is it that for those who now get to write and represent feminist (and) cultural theory class is not experienced or felt as immediately as gender? It may not be recognised as a problem for those who have the privilege to ignore it" (Skeggs 1997, 6).

In response to these (valid) points, I need to justify my exclusion of class, as both a construct conceptualised as relating to socio-economic status, or understood as more critically as relating to the social relations of production and distribution of material resources in society (Hartsock 1987). I had planned to incorporate class in this research. I provided women in fishing who were involved in the initial interviews with a sheet seeking information about their financial situation. Some women felt that the complexity and variability of their financial situation and ownership status made it impossible to accurately complete this sheet. Several women involved in this piloting stage also declined to complete the sheet or to answer questions about their financial situation and business arrangements. This could at least partially be attributed to the nature of the fishing industry, which is characterised by substantial competition and secrecy between fishers. Furthermore, business structures often involve multiple individuals, and some women may have felt uncomfortable providing information without the consent of their partners, husbands or other extended family members involved in the business.

A further factor compounding the complexity of incorporating class was the shifting fortunes between and within various fisheries. For example, the Western Australian Rock Lobster fishery was established primarily by European immigrants who typically had no economic or social status, and initially generated poor returns. However, there has been a shift in intervening years which has seen this fishery emerge as

Australia's most lucrative. This has resulted in particular social practices which complicate the operationalisation of class. For example, several older women in established multi million dollar, European extended family fishing businesses lived in palatial homes, with sports cars in their drive ways and expensive furniture, home entertainment equipment and household fittings. These women typically had four years of secondary education with many not having been engaged in paid employment, and were sending, or had sent, their children to expensive private schools. Yet these women were sceptical of tertiary education, and were encouraging their sons into their businesses and encouraging their daughters to seek "clean" work and then marry fishermen.

Conversely, some women sharing the similar social position of ethnicity and involved in similar business structures lived in modest homes, were engaged in paid employment and encouraged their children to seek tertiary qualifications. I also visited one modest home where a non European fishing family, who leased their rock lobster license, was struggling to make a living. This woman was tertiary educated, working in a poorly paid job in a rural area, had both her children at university and informed me that they were not allowed to go into the fishery. I felt these contradictions of experience, and the diversity amongst women reflected a complexity of the social position of class, which made it impossible to accurately incorporate this *within the scope of this research.* I was also reluctant to, and conscious of, inscribing a class position *onto* participants with insufficient exploration. Therefore, while I have not included class as a social location, far from dismissing the realities of these differences, I believe that, particularly in view of my findings, there is an urgent need to explore class as a social location in fisheries. Perhaps in future work, given the complexities of applying class in this context, employing the construct of status may be useful, as well as the application of a critical conceptualisation of class which explores the social relations of production and the distribution of material resources (Hartsock 1987).

Having justified my broad theoretical platform and the social locations addressed in this research, I now turn to specific concepts critical for understanding the contributions of women in fishing and their positioning in industry discourse. Applying post structural notions of language, subjectivity, discourse, power and knowledge are useful for understanding experience, and for analysing how gender relations are constituted, perpetuated and contested (Weedon 1989). In doing so, I draw from Foucault's conceptualisations of power, knowledge and discourse.

However, while many feminists acknowledge that these constructs inform a more sophisticated analysis (Weedon 1989) there have also been feminist critiques of Foucault's work. In this section, I explore these issues and explain my interpretation of theoretical constructs drawing from both Foucault and the feminist critique.

Theoretical guideposts-critical constructs

Language, subjectivities and subject positions

Language is central to any post structural analysis. I employ the concept of language as a discursive site, where subjectivities (a term often used interchangeably with the term "identities" in post structural writing) and social organisations are "defined, contested and constructed" (Weiner 1993). This exposes subjectivities as fluid and socially constructed, in contrast to a liberal humanist view of the individual, where the individual is the originator of his or her thoughts and speech (Weiner 1993). As Weedon (1989) explains, "language is not the expression of unique individuality: it constructs the individual's subjectivity in ways which are socially specific" (21). Therefore, meaning is constructed *within* language. Furthermore, as meaning is not guaranteed by the author's intent, linguistic analysis must consider interpretation and intent or impact (Weiner 1993).

In order to clarify the meaning of specific terms used in the book, I draw from Skeggs's (1997) interpretations of particular constructs. I apply the term "subjectivity" to mean "…the conditions of being subjected to frameworks of regulation, knowledge and discourse and constructing subjectivity in the process" (Skeggs 1997, 12). I employ the concept of "subject positions" to

> "…investigate the specificities of how women become particular subjects, especially respectable subjects. Subject positions are the effects of discourse *and* (organizational) structures. They are part of wider discourses" (Skeggs 1997, 12).

I use the concept of "positioning" to refer to how individuals are defined in and by particular subject positions, such as a "good wife".

Discourse

Post structural theorists view "truth"-ideas which have become "normal" and "natural" in specific historical and local contexts-"as problematic and open to scrutiny" (Weiner 1993). Discourse is the central principle of Foucault's genealogical method, where the exercising of power produces knowledge as "truth", and power is constituted in discourse. Foucault (1980) asserts that "...we cannot exercise power except through the production of truth" (93). Discourses are constituted in and through social institutions and individual subjectivities. Discourses are

> "...practices that systematically form the objects of which they speak...Discourses are not about objects; they do not identify objects, they constitute them and in the process of doing so conceal their own invention" (Foucault 1980, 49).

In this research I apply this notion of discourse, and attempt to scrutinise specific discursive text and practices to reveal how and why certain discourses appear more frequently, at this point in time, than other discourses.

Power/ knowledge and truth

In Foucauldian post structuralism, power is capillary, multiple and circulates; and is constituted and embodied in all relations and subjectivities. Power is not possessed in a quantity but *exercised.*

> "Power must be analysed as something which circulates....It is never localised here or there, never in anybody's hands, never appropriated as a commodity or piece of wealth. Power is employed and exercised through a net-like organisation. And not only do individuals circulate between its threads; they are always in the position of simultaneously undergoing and exercising this power....The individual is an effect of power, and at the same time, or precisely to the extent to which is that effect, it is the element of its articulation. The individual which power has constituted is at the same time its vehicle" (Foucault 1980, 98).

Furthermore, resistance is also central to the nature of power, and individuals demonstrate resistance in their daily lives by contesting subject positions.

> "… there are no relations of power without resistances; the latter are all the more real and effective because they are formed right at the point where

> relations of power are exercised; resistance to power does not have to come from elsewhere to be real, nor is it inexorably frustrated through being the compatriot of power. It exists all the more by being in the same place as power; hence, like power, resistance is multiple…" (Foucault 1980, 142).

Power/knowledge is conceptualised as inseparable and singular.

> "The exercise of power perpetually creates knowledge and, conversely, knowledge constantly induces the effects of power….Modern humanism is therefore mistaken in drawing this line between knowledge and power. Knowledge and power are integrated with one another…. It is not possible for power to be exercised without knowledge, it is impossible for knowledge not to engender power" (Foucault 1980, 52)

Thus, the production of "true" knowledge is inextricably intertwined with the power which produces and perpetuates it (Weiner 1993). Power and knowledge create "truth" as a performative mechanism which is constituted in and by dominant discourses.

> "There can be no possible exercise of power without a certain economy of discourses of truth .... We are subjected to the production of truth through power and we cannot exercise power except through the production of truth….Power never ceases its interrogation, its inquisition, its registration of truth: it institutionalises, professionalises and rewards its pursuit" (Foucault 1980, 93).

This exposes the political nature of all which seems apolitical, and focuses attention away from the content of discourse and towards the specific power dynamics inherent in discourse (Phoca and Wright 1999). Indeed, discourses are most effective when they are seemingly positive and apolitical (Phoca and Wright 1999). Power cannot be described as the result of one entity oppressing another, but it operates to uphold "truth" in the dominant discourse and *subjugates other knowledges* (Foucault 1980). Thus, bodies of "true" knowledge which account for the "normal" and "natural" at any one time appear in specific contexts, and who controls the agenda at any time is of paramount importance.

The "natural" and "normal"

Fixed "natural" and "normal" categories are constructed within the historical development of knowledges which reveal "truth". Thus, what we may see as appropriate behaviour for a "wife" may seem normal and natural but is infused with power. "Truth" maintains and reproduces the political and social world through regulatory understandings, such as what it is to be a "good" wife (normal and natural) or "bad" wife (abnormal and unnatural), which in turn legitimises this knowledge as "true". These "natural" categories operate as normalising regulatory discursive mechanisms to reproduce the existing social order, and explain how individuals come to adopt or reject certain subject positions and engage in, or do not engage in, particular activities. In applying this concept, I observe the following assertion by Weedon (1989) that there are particular "truths" associated with being a woman.

> "As women we have a range of possibilities. In theory almost every walk of life is open to us, but all the possibilities which we share with men involve accepting, negotiating or rejecting what is constantly being offered to us as our primary role–that of wife and mother. Whatever else we do, we should be attractive and desirable to men, and, ideally, our sexuality should be given to one man and our emotional energy directed at him and the children of the marriage" (Weedon 1989, 3).

As discussed previously, it is the relational oppositional construction of these fixed "natural" categories within particular discourses which enables one category to be construed and defined as superior to another category. Masculinity and femininity are relevant examples of this relational and hierarchical construction noted by many authors (eg, Bourdieu 2001; Connell 1995). The reinforcement of particular (traditional) masculinities in specific contexts is dependent on the subordination and derogatory construction of other categories, for example, "homosexual man" and "woman", which occurs through language, and reinforcing practices such as collective drinking (Gough and Edwards 1998; Willot and Griffin 1997).

> "The construction of devalued femininities thus provides an 'other' against which hegemonic masculinities constitute themselves" (Gough and Edwards 1998, 422).

Furthermore, language which incorporates *the physical* (eg, the size of breasts) into constructions of femininity, strengthens the association between "femininity" and "woman" (Gough and Edwards 1998).

> "By referring to women (im)purely in terms of disembodied features, the men are 'doing' their masculinity by reproducing sexist discourse....The use of language helps to alienate femininity by constructing it purely in the physical, thereby equating the feminine with female and hence marking

a strong distinction between feminine and masculine (male)" (Gough and Edwards 1998, 423)

Power as productive

Having explored how individuals are located within and in relation to certain subject positions, some discussion is required as to why people collude with, or resist, these subject positions. This raises Foucault's (1980) conceptualisation of power as not only coercive, but *productive*.

> "What makes power hold good, what makes it accepted, is simply the fact that it doesn't only weigh on us as a force that says no, but that it traverses and produces things, it induces pleasure, forms knowledge, produces discourse. It needs to be considered as a productive network ...much more than as a negative punishment whose function is repression" (Foucault 1980, 119).

As such, individuals willingly locate themselves in specific subject positions, which are productive of identity. Therefore, the system is self regulating and there is no need to ensure compliance through overt domination or force-discourse regulates in a self legitimising manner (Phoca and Wright 1999).

Feminists on Foucault

Many feminists recognise the benefits of applying a Foucauldian framework (eg, Bailey 1993; Cain 1993; Deveaux 1994; Grimshaw 1993; Ramazanoglu 1993). However, there have also been substantial feminist critiques of Foucault (eg, Bailey 1993; Deveaux 1994; Grimshaw 1993; Ramazanoglu 1993; Soper 1993). Foucault is charged with failing to recognise the systemic and pervasive nature of sexual privileging (Beasley 1999; Ramazanoglu and Holland 1993); and for failing to explore how individuals *experience and exercise* power (Phoca and Wright 1999). As Deveaux (1994) notes, there is "…the tendency of a Foucauldian conceptualization of the subject to erase women's specific experiences with power" (237).

These issues have significant implications for the politics of feminism and the empowerment of women (Deveaux 1994; Grimshaw 1993; Ransom 1993). Deveaux (1994) suggests that the lack of attention given to the *experiences* of individuals and how they inhabit their identities, particularly collectively,

> "...negates the importance of personal and group definition and affirmation, resources not easily replaced by the vague notion of identity contestation" (250).

Foucault is critiqued for failing to recognise notions of agency and empowerment by not exploring subjects' own understanding of oppression (Deveaux 1994). This lack of recognition of the personal experience of empowerment does not sufficiently recognise the essential nature of personal experience for change. Previously, feminism conceptualised women's liberation occurring through a process where women gain power *as a result of* gaining knowledge (McNeil 1993; Ramazanoglu and Holland 1993). In this context, it has been suggested that Foucault's conceptualisation of knowledge/power and self undermines feminist emancipatory claims (Ramazanoglu 1993) and that his agnostic conceptualisation of power provides neither an adequate understanding of empowerment and freedom, nor a framework to conceptualise empowerment (Deveaux 1994). Ramazanoglu and Holland (1993) state that

> "Feminism has preserved a more active conception of resistance linked to empowerment with which to challenge dominant discourses of sexuality" (258).

Foucault is charged with failing to offer any liberationary politics of resistance or change (Phoca and Wright 1999), including a lack of recognition of essentialism as a political strategy.

Thus, in addition to external sources of empowerment and disempowerment, it is also necessary to consider experience and internal sources that are fundamental to resistance and empowerment. Examples of internal sources of empowerment include love, and examples of internal sources of disempowerment include emotional labour, nurturing, care giving and shame (Deveaux 1994).

> "Addressing women's freedom requires that we reflect upon internal impediments to exercising choice as well as the tangible obstacles to its realization-and this means considering practices and conventions that may have disempowering effects not easily discernibile to theorists who focus exclusively on political power. Finally, it involves recognizing certain experiences as ongoing expression of resistance to power..." (Deveaux 1994, 246)

It is suggested that it is useful to conceptualise empowerment as a change in consciousness that results from the influence of internal and external factors (Deveaux 1994). In this framework, by positioning experience as a central feature, the process of empowerment relies (at least to some degree) on the conceptualisation of a self-determining individual. While this concept is fundamentally at odds with the underpinning tenets of post structural theory, it does enable us to ask

> "...what do relationships of power feel like from the inside, where are the possibilities for resistance, and what personal and collective processes will take us there?" (Deveaux 1994, 253).

Given that a rigid post structural framework forecloses the possibility that freedom can be achieved, a conceptual transition is needed from thinking of emancipating women from oppression to transforming existing political relations through production of new discourses (Ramazanoglu 1993). Ransom (1993) provides a positive reading whereby Foucault does not undermine agency, but centralises subjectivity by viewing it as a site contested in discourse.

> "So, for example, discursive constructions of the perfect mother exist, but are challenged by competing feminist conceptions of what women can be or of the ways women can be mothers. Women's identity as mothers is contested; a woman can resist the traditional discourse of motherhood by refusing to be a mother, by setting up alternative parenting arrangements, by taking her children on 'reclaim the night' marches or doing an evening course in roofing" (Ransom 1993, 134).

Similarly, Grace and Lennie (1998) argue that exposure to alternate discourses has implications for agency in possibilities for the (re)construction of women's identities; and found that the process of choosing between multiple contradictory subjectivities enabled women to become engaged in political activism.

I agree with the assertions by Deveaux (1994), that it is incorrect to read Foucault as a relativist (as, for me, the concept subjugated and dominant knowledges imply a *limited* plurality), or assert that he does not recognise domination. Several aspects of Foucault's work are indicative of this last point and, of particular relevance to this research are his reflections on the state as "meta power".

> "The State is superstructural in relation to a whole series of power networks that invest the body, sexuality, the family, kinship, technology, and so forth…this meta-power with its prohibitions can only take hold and secure its footing in a whole series of multiple and indefinite power

> relations that supply the necessary basis for the great negative forms of power" (Foucault 1980, 122).

Applying the notion that power is productive, the power to produce *control* is contingent on the state's ability to subversively engage the participation of individuals (Phoca and Wright 1999). The political nature of Foucault's work is also apparent in his concern with the

> "...*effects* of the centralising powers which are linked to the institution and functioning of an organised scientific discourse within a society such as ours" (Foucault 1980, 84, italics added).

Reading gender into this statement, I also note the masculine connotations inherent in an "organised scientific discourse". Furthermore, Foucault is concerned with the "insurrection" of subjugated knowledges, which are revealed (as opposed to knowledge not yet gained), and which emerge through a critical interrogation of discourse.

> "Subjugated knowledges are thus those blocs of historical knowledge which were present but disguised within the body of functionalist and systematising theory and which criticism...has been able to reveal.... a whole set of knowledges that have been disqualified as inadequate to their task or insufficiently elaborated: naive knowledges, located low down on the hierarchy, beneath the required level of cognition or scientificity" (Foucault 1980, 82).

Foucault (1980) also asserts that

> "To say that one can never be 'outside' power does not mean that one is trapped and condemned to defeat no matter what" (141-2).

Conclusion

Thus, while applying Foucauldian concepts in this research, I expand these conceptualisations by incorporating the observation that experience, emotions and internal processes (eg, shame, love and reputation) also influence women's subjectivities and discursive positioning. Employing this approach, which draws from both Foucault and the feminist critique of his work, is consistent with my application and understanding of the partial and situated nature of knowledge (Haraway 1988). Employing this theoretical approach enables me to recognise the diversity of women in fishing while simultaneously speaking of women's collective experiences and knowledge. This allows me to retain the feminist political integrity necessary to further the cause of women in fishing.

In the next chapter, The Doing of Feminism, I discuss the feminist methodological principles underpinning the research and overview the research process.

CHAPTER THREE

THE "DOING" OF FEMINISM

Introduction

In this chapter, I explain the feminist methodology employed in this research. Feminist epistemological and methodological issues are explored to increase the rigour of this research and counter the lack of methodological discussion in feminist rural sociology (Pini 2003a). The principles which inform this feminist research are articulated, the criteria for evaluating feminist research discussed and feminist concepts of validity and generalisation explored. Each section also explains how these concepts and principles were applied throughout this research. I then present an explanation of, and justification for, the research design and process.

Methods, Epistemology and Methodology

In the past two decades, there has been increasingly sophisticated discussion on feminist research and methodology (Fawcett et al 2000; Fonow and Cook 1991; Harding 1987a, 2004a; Hesse-Biber, Gilmartin and Lyndenberg 1999; Nielsen 1990; Reinharz 1992, 1997). Not only has this discussion centred on the distinction between feminist and mainstream research, but on the nature of feminist research itself. When reviewing the literature it becomes apparent that there are as many approaches to feminist research and methodology as there are "feminisms" (Humm 1995). The practice of feminist research is also inextricably linked to theory and, indeed, is the "doing" of feminism (Kelly, Burton and Regan 1994, 28).

Harding (1987a) suggests that a useful approach to discussing feminist research is to distinguish between methods, epistemology and methodology. These respectively refer to the techniques used to collect data, the philosophical grounding for the research, and the theory and analysis of the research process. In order to clarify what distinguishes research as "feminist", I now provide a brief discussion on each of these concepts.

Defining feminist method

The question of whether there are, or should be, distinctive feminist methods has been addressed by many authors (eg. Harding 1987a; Reinharz 1992). Initially, second wave feminism aligned positivism with the dominant quantitative method, critiquing the latter on the absence of women, the inability to explore women's experiences and the prevalence of predetermined categories (Maynard 1994). These inadequacies, combined with the ontological and epistemological critique of positivism's "objectivity", led feminists to advocate qualitative methods as necessary for feminist research (Maynard 1994; Stanley and Wise 1993).

However, post structural theorists have since challenged the ultimate "truth" of all discourse, and deconstructed the binary categories of qualitative research/ feminism and quantitative research/ science. While some feminist scholars maintain that qualitative methods are the only appropriate methods for feminist research (eg Stanley and Wise 1993), quantitative methods can be valuable for increasing the potential political influence of feminist research (Gottfried 1996; Maynard 1994; Spalter-Roth and Hartmann 1996). In sum, a wide range of methods are utilised in feminist research and the increasing use of mixed method designs is recognised as a trend (Maynard 1994). Therefore, it is not the methods employed which make research "feminist", but the underlying epistemological and methodological principles (Ardovini-Brooker 2001; Harding 1987a). Indeed,

> "the political nature of feminist intellectual practice demands that feminist method works as strategy rather than as some stable notion of technique" (Waldby 1995, 28).

To further explore the making of research "feminist", we must address notions of feminist epistemology.

Feminist epistemology

Associated with the Enlightenment and modernity, traditional epistemology is based on positivist standards of objectivity and "truth" as idealised in the physical sciences. The feminist critique, which "…spat in its eye" (Stanley and Wise 1993, 66), has focused on issues of power and position, including the exclusion of women as "'knowers' or *agents of knowledge*" (Harding 1987a, 3), the underlying gendered politics of knowledge (Logan and Huntley 2002), and the critique of objectivity (Bhavnani 1993; Harding 1987b; Haraway 1988; Waldby 1995). Contrary to the "add women and stir" approach often evident in mainstream research, feminist epistemology proposes "...alternative theories of knowledge that legitimate women as knowers" (Harding 1987a, 3).

Three central feminist epistemologies are typically discussed-empiricist, standpoint and post modern epistemology (see Harding 1987b for a detailed discussion). These are not mutually exclusive-not one alone resolves all feminist debates-and differing philosophies of science and "feminisms" inform epistemologies, which lead to the production of different kinds of analysis and knowledge (Campbell and Wasco 2000). This diversity in claims to knowledge production validates the sentiment that feminist research is not about replacing the dominant positivist tradition with a singular feminist strategy for eliciting truth (Ardovini-Brooker 2001; Harding 1987b).

Feminist empiricism focuses on addressing the androcentric nature of "bad science". Within this framework both qualitative and quantitative methods are used. However, there is an emphasis on intersubjectivity and commonly created meanings, on recognition that bias is introduced in the research process and that the knowledge produced is influenced by the social identity of the researcher (Harding 1987b).

Standpoint epistemology locates the problem as beyond "bad science". It is argued that the specific social location of the knower is important because it shapes what is known and that women's experience (or the view from below), when informed by feminism, enables more accurate knowledge claims than men's (Harding 1987b; Hartsock 1987). Women researchers may be seen as "outsiders within", as they share oppression with women yet are privileged by their researcher status (Collins 1991). Research must begin from women's lives, i.e. the view from below, as women's location is a resource for the construction of new knowledge (Harding 1987b). However, experience is recognised as problematic and as systems of oppression also shape the consciousness of the oppressed, knowledge that reflects the experiences of both dominant and subordinate groups must be constructed.

Assertions include: that women's experiences are shaped by everyday material experience and thus men's and women's experiences differ; that privilege is least visible to those who control the resources that define dominant ideology; and that it takes the standpoints of the oppressed to recognise systems of oppression and privilege (Harding 1987b). While standpoint theorists differentiate standpoints based on social positions such as race and class, it can be argued that both standpoint epistemology and feminist empiricism conceptualise men and women as "sex classes" (Harding 1987a). As such, standpoint theorists have been charged with essentialism and of repeating the error of androcentric science in privileging the view from below.

Empiricism and standpoint epistemology derive respectively from liberal and socialist feminist theory, which are concerned with modernist emancipatory aims. These epistemologies aspire to enhance the objectivity of science (Harding 1987b), and are imbued with the notion that overcoming oppression is possible (Fawcett and Featherstone 2000). In contrast, post modernist feminist epistemology is based on a relativist position, which asserts that a structural perspective incorrectly assumes the ability to produce one true story about women's position in society-including an aspiration to freedom-and instead asserts that all stories are equally valid. However, conceptualising feminist knowledge in relation to relativism is challenging given feminism's political agenda (Fawcett and Featherstone 2000).

"Situated knowledge"

These tensions in feminist epistemological debate relating to "truth" or the privileging of knowledge, and conceptualising difference while retaining unity, have led to the development of alternate feminist epistemologies that enable feminist praxis (Gorelick 1996; Haraway 1988; Stanley and Wise 1993). Haraway (1988) draws from both post modern relativism and standpoint theory to develop her concept of "situated knowledge" which recognises all knowledge as partial and situated. The advantage of her conceptualisation is that a standpoint is not understood as starting from women but as starting from difference.

I have adopted "situated knowledge" (Haraway 1988) as the feminist epistemology informing this research, as it allows both difference and unity to be conceptualised in a framework that legitimates women as knowers, enables empirical research to occur and generalisations to be made. Ontologically, this view equates with a social constructionist understanding where "...collective sense is made, meanings negotiated and identities elaborated, through the processes of social interaction between people" (Kitzinger 1999, 65). To understand how this feminist epistemology is transformed into feminist practice we must turn to feminist methodology.

Feminist methodology

Feminist epistemology informs feminist methodology. However, the diversity of feminist epistemological foundations renders a singular definition of feminist methodology impracticable, with many authors instead discussing common themes or principles of feminist research (Armstead 1995; Bhavnani 1993; Cook and Fonow 1990; Reinharz 1992; Shields and Dervin 1993).

In this research, I employ the five common themes which characterise feminist research extrapolated by Pini (2003) in her review of the feminist methodological literature-a focus on gender, value given to women's experiences and knowledge, rejection of the split between subject and object, emphasis on empowerment and emphasis on political change/emancipation. As Pini (2003) observes, these principles are interdependent, not exclusively feminist, change with developments in feminist theory and each one is subject to debate. I complement these guiding principles by employing Bhavnani's (1993) model for evaluating feminist research (or achieving feminist objectivity) where three pragmatic criteria-micropolitics, reinscription and difference-derive directly from Haraway's (1988) constructs of partial knowledge, accountability and situated knowledge. These concepts are easily incorporated within Pini's (2003) five principles. A further appeal in employing Bhavnani's (1993) criteria is that she translates her three concepts into concise questions that may be used to evaluate feminist research. I now provide a more detailed explanation of these concepts and a discussion on their application in this research.

Feminist Research Principles, Application and Evaluation

A focus on gender

Research informed by feminist epistemology, which legitimates women as knowers and prioritises women's voice and visibility, has gender at its centre. Demonstrating a focus on gender, this research involved engaging 65 women in fishing in interviews or focus groups. Pini (2003) identifies two important issues relating to a focus on gender-firstly, the place of men and masculinities in feminist research and secondly, accommodating diversity and difference among women.

Men and masculinities

As Pini (2003) notes, much feminist scholarship advocates for the study of men and masculinities in the context of a critical analysis of gendered power relations that are identified as requiring change. Furthermore, she argues that recent deconstructive scholarship on agricultural discourses, which explores how gendered identities are powered and positioned by Brandth and Haugen (2000) and Liepins (2000), represents a major advance in this area for rural sociology.

The domination of men in fisheries as license holders and position holders in industry organisations and fisheries management agencies requires me to justify my primary research focus on women in fishing. I chose not to interview men in fishing for several reasons. Firstly, fewer women would have been able to be interviewed, thus reducing the focus on giving voice to women in fishing, highlighting their contributions and increasing their visibility. Secondly, I was concerned that the research focus may subtly be appropriated away from the experiences of women in fishing and their contributions, to men's or family experiences and contributions. Thirdly, I believed that restricting the research focus to women in fishing may affirm the value of their contributions. Fourthly, I was concerned that women in fishing may censure their stories if their partners, or men they knew, were also participating in the research.

For these reasons, interviews were conducted primarily with women in fishing. However, to ensure that dimensions of gendered power were incorporated into the research, men and women in leadership positions in industry organisations and in fisheries management agencies were interviewed. Additionally, interviews with women in fishing included discussions about the recognition they received for their contributions from their partners and the broader industry. As a further measure to incorporate gendered power relations, cultural and historical resources pertaining to case studies were also analysed.

Difference and diversity

Accommodating difference and diversity amongst women is a central concern in feminist research. As Harding (1987a) notes, women's experiences vary both across cultures and within individuals in the form of contradictory fragmented identities-"...it is 'women's experiences' *in the plural* which provide the new resources for research" (Harding 1987a, 7). Bhavnani's (1993) first evaluative question for feminist research corresponds to Haraway's (1988) concept of partiality and relates to difference.

> "In what ways are questions of difference dealt with in the research study-in its design, conduct, write-up, and dissemination?" (Bhavnani 1993, 99)

This research accommodates difference at a theoretical level by employing Haraway's (1988) "situated knowledge", which recognises all knowledge as partial and situated. At a practical level, difference has been incorporated in this research in several ways. The case study approach was employed to involve women in fishing from varying geographic locations and fisheries, and to incorporate both macro and micro gendered social relations. Diversity was also a critical factor in the selection of case studies. Multiple techniques were employed to recruit women in fishing as potential participants. This approach ensured a diverse range of women was included in the research; and facilitated the involvement of less visible groups of women in fishing, including women with diverse ethnic backgrounds and women with small children. The use of purposive snowball sampling, including during fieldwork, and the process of screening women for the interview process on the basis of basic demographic information ensured a diverse range of women participated in the research. Semi structured interviewing also provided the opportunity to explore their diverse experiences.

During analysis and write up, I was committed to reflecting multiple and contradictory realities (Stake 1995). Furthermore, this book highlights barriers women experience and discusses strategies for more adequately incorporating women's diversity in industry practices and processes.

Historical dimensions

Bhavnani (1993) also asserts that incorporating the historical dimensions of social relations is essential for understanding difference. The case study method is particularly suited to this type of investigation. Incorporating historical dimensions was achieved through the analysis of available cultural and historical resources in case studies and through semi structured interviews, which enabled women to tell the stories of their industry involvement. Interviews also elicited information on the past, present and predicted future contributions of women in fishing.

A value given to women's experiences and knowledge

Valuing women's experiences and knowledge is central to feminist epistemology. My adoption of this principle was demonstrated by the number of women in fishing involved in the research, and through the research methodology and methods. While the importance of incorporating difference has been discussed above, other issues emerge in relation to valuing women's experiences and knowledge (Pini 2003a). Pini (2003a) critiques feminist rural scholarship on its lack of explanation with regard to *how* "giving voice" to women is achieved and argues that these processes need to be more clearly articulated. In addition, women's own perception of their experiences and knowledge needs to be considered. For example, Pini (2003) discusses how some women in her research doubted the value of their experiences and knowledge. Focusing on how voice is given also legitimates and allows voice to be given to the researcher's experiences (eg, Logan and Huntley 2002). In this process, reflexivity is a key concept, which allows a researcher to reflect on her or his subjectivity and positioning, and the impact on the research process (England 1994). A more thorough analysis of reflexivity and its application is provided in the discussion on the split between subject and object.

Giving voice

Voice was/is given to women in fishing in three aspects of this research-through the methods employed, the management of interpretations and the dissemination of findings. The use of semi structured interviews as a method of data collection provided the potential for issues to be fully explored, enabled women to express thoughts and feelings about their experiences and knowledge, and allowed women to raise and discuss issues they believed were important. In addition, employing purposive sampling allows for the inclusion of participants whose voices otherwise would not have been heard (Kitzinger and Barbour 1999).

The interpretation of voice was also considered. While several authors advocate providing participants with the prepublication draft to check that understanding and interpretation is congruent with their understanding, four issues relating to power and positioning influenced my decision not to do this. Firstly, practical limitations included geographical distance and the number of women involved in the research. How could I incorporate the comments of all of these women?

Secondly, conceptualising knowledge as situated and partial means there will always be difference between my understanding and the understanding of women in fishing. This is not to privilege my academic understanding above that of the experiential knowledge of women in fishing, but to acknowledge my understanding as also situated and partial. Previous research indicates that a participant's and researcher's relationship to feminism is likely to be one such point of divergence (Armstead 1995; Gorelick 1996). Research in Australia has found that women in farming and rural women are reluctant to identify with the construct and/or term "feminism" (James 1989; Franklin 1989; Haslam-McKenzie 1998; Hogan 1994; Grace and Lennie 1998; Teather 1996a, 1998). Alston (1994a) explains that women may reject feminism as it threatens the nature of their family economic unit, and asserts that a further issue in this regard is the urban bias of feminism (Alston 1994a).

Thirdly, unavoidably I retain power over the final written product/s and, ultimately, I still must choose to include/ exclude/ interpret differing understandings. This is particularly pertinent in the context of my belief that giving voice alone is insufficient, and that interpretation and analysis is required if outcomes are to emerge from research. Fourthly, women may be reluctant to comment honestly on the draft (Armstead 1995).

Despite these issues, I was conscious to avoid a total appropriation of women's voice. I encouraged women in fishing to reflect on their involvement in the research and to contact me to ask questions or add further comments. Direct quotations were included in the write up and a commitment made to expressing divergent viewpoints. In addition, women were asked to comment on the practices and processes of industry organisations and fisheries management agencies, it is hoped that the publication of this book will have a real impact on these practices and processes.

Rejection of the split between subject and object (or feminist objectivity)

The feminist critique of the "disembodied scholar" (Waldby 1995) has rejected notions of impartial objectivity, and is concerned with the interrelated issues of positioning and the critical analysis of power (Harding 1987a; Stanley and Wise 1993). The "object" of research may be articulated not as "...the unmediated world of the researched, but on the world *between* ourselves and the researched" (England 1994, 87).

Many issues arise in relation to this feminist critique. The research process is dialogic and personal as both the researcher and participants structure the research (England 1994; Stacey 1988) and research is influenced by emotions (Stanley and Wise 1993). The hierarchical ordering of the researcher as "expert" is inverted and the "view from below" adopted (Mies 1983, 123) with knowledge claims voiced from a position of "explicit uncertainty" (Waldby 1995, 27). Many researchers have an aversion to "mining" participants for information (Stanley and Wise 1993) and aim for reciprocal democratic relationships based on principles of empathy and respect (England 1994) and the sharing of skills and information (Pini 2003a).

However, even in these exchanges, researchers may not reveal certain aspects of their lives to participants (Armstead 1995). The perception of power dynamics may differ. While, researchers may assume they are more powerful due to their position as researcher/ academic/ feminist, participants may feel these positions are worthless (Armstead 1995). Women interviewing high profile women and men may be relatively less powerful or intimidated in an exchange (Acker, Barry and Esseveld 1996) and researchers may be ridiculed if they identify as feminists (Armstead 1995). While power ultimately rests with the researcher's responsibility for constructing the finalised research product, this power may be negotiated or constrained by funding organisations, research teams and institutionalised, androcentric sociological practices (Smith 1987).

Therefore, it is important for researchers to expose the partiality of their perspective. Researchers

> "...need to locate ourselves in our work and to reflect on how our location influences the questions we ask, how we conduct our research and how we write our research" (England 1994, 87).

Reflexivity is central to Bhavnani's (1993) second construct of micropolitics, which corresponds to Haraway's (1988) concept of positioning. Bhavnani's (1993) concept of micropolitics enables her to frame the second question by which feminist research may be evaluated.

> "How and to what extent does the research conduct, write-up, and dissemination deal with the micropolitics of the research encounter-what are the relationships of domination and subordination which the researcher has negotiated and what are the means through which they are discussed in the research report?" (Bhavnani 1993, 98).

Reflexivity

Reflexivity is a strategy utilised by many feminist researchers because it is seen as the key to producing less distorted knowledge (England 1994; Harding 1987a). This concept defies the notion of an impersonal objective researcher and refers to the "…self-critical sympathetic introspection and the self-conscious *analytical* scrutiny of the self as researcher" (England 1994, 82). Thus the biography of the researcher becomes central, as placing oneself on the "same critical plane" (Harding 1987a, 9) allows for information about a researcher's beliefs and social position to be considered in the context of the overall research frame. Researchers therefore appear

> "…not as an invisible, anonymous voice of authority, but as a real, historical individual with concrete, specific desires and interests….We need to avoid the 'objectivist' stance that attempts to make the researcher's cultural beliefs and practices invisible while simultaneously skewering the research objects beliefs and cultural practices to the display boards" (Harding 1987a, 9).

While reflexivity may increase awareness of the positioning of the researcher, it does not *eliminate* power dimensions (England 1994). Pini (2003) asserts that, corresponding to the lack of attention given to methodology in rural feminist research, there is an oversimplification of the participant/ researcher relationship, where the sharing of a common characteristic, often gender congruency, is seen as providing the basis for a shared understanding. This oversimplification may also occur where researchers aim for a reciprocal research process (Shields and Dervin 1993). In both cases, what often fails to be disclosed is the power differential between the researcher and the researched, and the recognition that partiality and situation limit understanding, which may mask exploitation (Armstead 1995; England 1994; Mies 1991; Pini 2003a). Despite shared understandings between the researcher and the researched, the researcher remains positioned as a researcher (Acker, Barry and Esseveld 1996).

Locating myself in the research process

Throughout the research process, I maintained a journal where my reflections on power and positioning enhanced my understanding and interpretation of power relations in my research. Furthermore, in an effort to address my own lack of visibility and to demonstrate my own historical, concrete person (Harding 1987a), I have also included a brief biography in the Introduction.

Positioning

My experience, knowledge, qualities and skills fostered my ability to engage with feminist research practice in particular ways. I developed rapport with many women in fishing by relating through differing aspects of identity within my subjectivity such as researcher, student, feminist, teacher, learner, mother, as wife and single. While interviewing women with small children, I nursed babies, assisted with colouring in and discussed childhood development issues. While interviewing women who were participating because they believed passionately in the need for increasing recognition for women in fishing, I found myself drawing more heavily on feminist aspects of my subjectivity. Reflecting the influence of the research process on both the researched and the researcher, in three interactions in particular, I was humbled by the adversity, honesty and courage expressed in women's stories. Interviews characterised by a high degree of congruence of, and connection around, specific aspects of identity shared between myself and participants were fluid, engaging, humorous and sad, and followed the interview schedule less rigorously.

While I felt rapport with many women at different points during interactions, there were also many moments where my values and beliefs diverged significantly from those of the women I was interacting with. This occurred most commonly in relation to women's understanding of the value of their own contributions, and in relation to the derogatory interpretation by some women in fishing of alternate feminist discursive scripts.

In these instances, my identity as researcher became more prevalent. These acknowledgements again highlight the partial and situated nature of knowledge including mine, the readers' and the women involved in this research. There were also some women I felt unable to engage with for no identifiable reason and these exchanges were more formal and impersonal. These exchanges were typical in my interviews with women in Cardea. Factors which could account for this may include my being viewed as an "outsider" by women in this close knit Italian community, my own lack of cultural awareness, and the fact that these women were related to each other.

The practicalities of case study fieldwork also influenced my positioning in the research. The wide geographical scope of the research, combined with my family responsibilities, meant that my fieldwork time was limited and pre-planned to a great degree. In addition, I had to fit in with women's personal schedules. This meant that, on most days, I conducted between three and five interviews, with some interviews conducted in the evening. On one day in particular, my research journal reflects my feelings of fatigue and the acknowledgement that I could have conducted a particular interview in a more focused manner. In addition, the boundary between where research began and finished was also unclear in several instances. I had previously met one participant on several occasions, arranged to interview her and, as she lived in a remote location, accepted her invitation to stay the evening at her home. After dinner, she discussed some highly personal issues. This raised the ethical dilemma about what was included in the case study research, and what was outside the research (Gorelick 1996). In this instance, while our discussion could be considered relevant to the research, I chose not to include information disclosed to me after the formal interview was concluded.

I employed numerous strategies throughout the research to address the issues of positioning and to reflect my feminist commitment to rejecting the spilt between object and subject. These included: entering into a process of industry immersion; generating discussion and receiving input into the research; sharing my knowledge of industry organisation and fisheries management agencies' decision making processes; sharing knowledge about WINSC where women were interested; encouraging women throughout the research process to contact me to ask questions or access information; employing semi structured interviews as a research method to allow for more conversational, empathetic interaction; employing focus groups as a less hierarchical research strategy; collaborating with women about possible outcomes; becoming a financial member of WINSC; and employing a gendered "I" in the research as a formal methodological intervention to locate myself in the research (Waldby 1995).

Power

Issues of positioning relate to the power dynamics in the research process. The communication style of individual women influenced my need as researcher to exercise power during the interview process. For example, while some women provided short concise answers to each question, other women provided more lengthy answers which would deviate from the question. On these latter occasions, I exercised power through my positioning as researcher by refocusing the interview and choosing which issues women raised would be explored, and the degree to which these issues were discussed. In addition, where women had limited time to participate, I was conscious of my power in selecting the topics to be explored. Several participants, in expressing admiration for my educational attainment, researcher status and assumed knowledge, questioned the value of their participation and reflected their own positioning as secondary. While my position as researcher was inverted when women asked questions of me, at all times I retained power over the type and amount of information I shared about myself and how and when this occurred.

Power dynamics were also inverted at other times during the research process. One woman dominated the interview through her communication style and, during this interaction, I felt unable to refocus and direct the process. Some women perceived the research as a waste of time and were sceptical that the research would result in any practical outcomes. During these exchanges I experienced a sense of powerlessness and inadequacy relating to the spilt between women in academia and women's activism (Smith 1996), as I could not guarantee that these individual women would benefit directly in a tangible way from the research. In several interviews with men who held positions in industry organisations or fisheries management agencies, I also sensed a degree of ridicule and scepticism about attributing value to the contributions made by women in fishing.

While institutions and organisations may influence research design, outcomes and uses (Gottfried 1996; Smith 1987), I experienced support from the Bureau of Rural Sciences (BRS) and Charles Sturt University

(CSU) for employing a feminist approach. As Pini (2003) notes, researchers can be charged with political bias and consequently not have their research taken seriously, or be held responsible for the problem under investigation. While presenting on the research during a visit to Agriculture, Fisheries and Forestry Australia (AFFA), I was challenged about the exclusion of men from the research and then asked how I could ensure the equitable distribution of fishing rights for women. These exchanges initially left me with a sense of powerlessness and intimidation due to the high profile positions these individuals held and my lack of experience with, and knowledge of, the fisheries management agencies. However, as the research progressed this feeling was replaced by a sense of hope (power) that, potentially, I may be in a position to influence this gendered thinking.

Emphasis on empowerment and political change/ emancipation

Empowering women and seeking broader political change/ emancipation is a common aim and method of feminist research. Hartsock (1987) argues that politically focused research provides a less distorted view as oppression can only be overcome by struggling against and revealing the concealed social relations that represent exploitative systems. Feminist research is politically motivated or designed "…*for* women" with the goal of providing explanations that women want or need (Harding 1987a, 8).

"Consciousness raising", where a researcher facilitates group participants understandings of their social positions, has historically been a feminist strategy employed to empower women (Reinharz 1992; Shields and Dervin 1993). However, a more sophisticated analysis of power in the research process raises concerns about judging the validity of women's understandings of their experience and knowledge, and the privileging of knowledge. Pini (2003) provides an alternate conceptualisation of empowering women by employing post structural theory which challenges the liberationist discourses of feminism, and in which subjectivities are fluid and may be positioned in multiple discursive scripts. Pini (2003) retains her commitment to emancipate and empower women but locates this "within the lived experiences of the women" she conducted research with (117). Leipins (1998), also adopting a post structuralist position, argues that providing the opportunity for discussing gender issues allows women to choose between contradictory subjectivities, thereby facilitating the potential to demonstrate political agency.

Facilitating opportunities for the empowerment of women in fishing was a research objective and women who participated were encouraged to ask questions, seek further information and discuss the research. I discussed gender with women in fishing who were interested in this issue, thereby exposing these women to an alternative and conflicting discourse and facilitating the potential for altered subjectivities. Focus groups proved to be a powerful mechanism in this regard which allowed contradictory subjectivities to be disclosed and discussed. Some women involved in focus groups reflected that participating had altered their views of the industry. Several participants expressed a new found interest in political action and a desire to "be heard" by fisheries management agencies and industry associations. With consent, I collected a list of names of women in fishing who wished to find out more about WINSC and forwarded their names to the organisation, and provided those interested women with a WINSC brochure.

Another strategy to empower women is to encourage them to control and provide input into the research process. This involvement ranges from the extensive involvement in action research (Mies 1983; 1991) to participation or collaboration in certain research aspects such as research design or commenting on draft reports. However, collaborative research can be demanding of participant's time and energy (Armstead 1995) and the degree of collaboration varies between projects, even when collaboration is employed as a specific methodological strategy (Reinharz 1992). Furthermore, as discussed above, the ultimate power to include/ exclude/ accept/ reject/ alter research lies with the researcher. However, by incorporating women's input into the research design and research questions and in exploring women's desired research outcomes, there was a degree of collaboration in this research.

Facilitating broader political action was also a research objective. Strategies adopted to increase awareness of the contributions of women in fishing and facilitate political action included: placing articles in industry magazines; providing all fisheries management agencies and key industry organisations with the opportunity to participate in the research; defending the feminist orientation of the research in several arenas; establishing my own industry network to increase the possibilities for action resulting from the research; adopting a flexible approach to enable opportunities for action to be taken up during the research process such as responding to requests from management agencies to provide advice; and the publication of this book.

The inherent appropriation in textual production (England 1994) and the construction of an alternate and competing discourse generated by the researcher (Jupp and Norris 1993, 51) also has implications for women's empowerment and political change. There are difficulties in documenting contradictory interpretations without imposing labels and reinscribing women into stereotypical roles, while simultaneously privileging neither feminist academic, nor individual experiential knowledge (Armstead 1995, 633).

> "Very subtle questions of constructing self and other through our written representations have important consequences for the power relations of research and the political goals of feminism....I feel that the political challenge is to address the contradictory truth that our writing often has little impact on the worlds of "others" which we describe" (Armstead 1995, 636).

These issues reflect the final construct of reinscription proposed by Bhavnani (1993) as an evaluative concept for feminist research, which she derives from Haraway's (1988) concept of accountability. Her corresponding question to evaluate feminist research is as follows.

> "Does this work/ analysis define the researched as either passive victims or as deviant? Does it reinscribe the researched into prevailing representations?" (Bhavnani 1993, 98).

Throughout the research process, I was committed to seeing, hearing and exploring the agency and resistance of women in fishing, while simultaneously understanding and locating this in the context of women's lived experience. Voice was given directly to women through the opportunity to use their own words to comment and/or question practices and processes of industry organisations and fisheries management agencies. This is seen as one mechanism to raise awareness of the contributions of women and to affirm, for women themselves, the value their opinions.

Furthermore, in addition to considering the maximum potential for political action, avoiding women's reinscription was an influential factor in selecting which information I presented in different arenas. For example, when presenting to policy officers in fisheries management agencies, I consciously focused on the extensive contributions of women in fishing, the structural barriers to their greater participation and strategies to enhance their participation. In addition, a post structural theoretical framework was employed to avoid the essentialist representation of women in fishing as structurally oppressed, and to facilitate recognition of the potential for women's subjectivities to shift as a result of exposure to competing discursive scripts. Furthermore, this book includes significant theoretical and methodological detail in order that its value may be assessed by researchers, academics and policy makers.

Feminist Validity and Generalisability

While feminist scholarship typically addresses objectivity, discussions of validity-the positivist concept referring to the accuracy of correspondence between construct and conclusion-are often absent in feminist methodology. Yet, feminists can counter common criticisms of their work and increase the rigor of their research by specifically addressing validity (Dallimore 2000).

Interpretive scholarship has questioned the relevance of a positivist conceptualisation of validity to social science and has proposed the alternate construction of validity as the "trustworthiness" (Dallimore 2000), or "adequacy" (Acker, Barry and Esseveld 1996), of particular interpretations. While adequacy or trustworthiness can not be absolute, this conceptualisation of validity reflects an increase in conventional corresponding notions of internal and content validity (Dallimore 2000). This can be achieved by: using a variety of methods and sources or "triangulation"; by using findings to achieve change for women-i.e. using practice as a measure for validity where a measurement technique is deemed successful if it facilitates a specified outcome (Dallimore 2000); ensuring women's voices are directly heard in the account (Acker, Barry and Esseveld 1996; Merriam 1998); and employing reflexive processes and collaborative research techniques (Acker, Barry and Esseveld 1996).

Dallimore (2000) also argues that feminist research must reconceptualise positivist notions of generalisability or external validity and draws on the interpretive paradigm's construction of generalisability as "applicability". She proposes adopting specificity and accountability as two methods by which feminist researchers can demonstrate the applicability or generalisability of their findings and overcome the essentialist nature of traditional generalisation. As Reinharz (1992) also notes, exploring "specifity, exceptions and completeness" (174) counters the focus of science on traditional generalisation which has rendered women historically invisible.

Dallimore's (2000) discussion of specificity draws on the work of Phelan (1994 cited in Dallimore 2000), who proposes the examination of specific similarities and differences as an approach to avoiding essentialising while maintaining the cohesiveness required for political action. Indeed, specificity, and the concept that the general lies in the particular, is well established in discussions of qualitative generalisability (Merriam 1998). Phelan (1994, cited in Dallimore 2000) argues that focusing on similarities and "weaving threads together" (164) where "threads" are conceptualised as aspects of identity which can not be separated, bridges difference, and that this specifity enables political action. However, this political action is not based on "identitarian formulations", which assumes similar identities, but is political action based on "alliances", or coalition politics based upon community. Dallimore (2000) argues that it is this specificity or "weaving together" that provides the foundation for bounded generalisation and that either reconceptualising generalisability in terms of specificity, or utilising bounded generalisation as a form of specificity, addresses both population and ecological validity.

> "In its bounded form, generalizability refers to people who share identity points of connection and who construct community through their shared political action. Specificity offers a bounded form of generalisability by which feminist researchers come to understand the specific experiences of participants' lives can then begin to relate their findings to others outside their immediate study who share similar "identity points" and who may be willing to come together to create a community through action and social change for women....Being bounded does not mean identifying a smaller population but rather a population with overlapping similarities. Further, specificity provides a process of social change consistent with feminist goals (while addressing concerns of essentialism) that ultimately allows researchers to make themselves accountable to their research participants" (Dallimore 2000, 165).

Reliability too can be reconceptualised as dependability, or the consistency between the data collected and the results (Merriam 1998, 206). This can be achieved through discussing the positioning of the researcher, triangulation, maintaining an audit trail or descriptions about design data collection and analysis, and providing justification for decisions made throughout the research process (Merriam 1998; Yin 2003a).

In this research, feminist validity is conceptualised as trustworthiness or adequacy of interpretation. Specificity is consistent with Haraway's

(1988) situated and partial knowledge. Bounded generalisation is adopted in this research as a form of specificity, where shared social positions provide the basis for these generalisations. Accountability and dependability are also adopted as guiding principles of the research. Strategies employed at a practical level to ensure feminist validity and dependability include: collaboration; triangulation of methods; the development of specific political actions determined in collaboration with participants; a transparent research process; and discussing power and positioning in the research.

Having established the nature of feminist research, explored the principles which guided the research and their application and discussed both the criteria for evaluating the research and issues of feminist validity, I now turn to a description of the research process employed in this research.

Research Design

This research was based on a three stage design. Stage One evolved from the original research proposal and informed the research. Stage Two involved research on women's contributions at a national level and Stage Three consisted of three case studies to ensure difference and diversity was adequately incorporated into the research.

Role of theory

Research may be understood as continually moving between a process of induction and deduction (Merriam 1998). The research questions were developed from theoretically informed literature that guided data collection throughout the research. Theory was generated from themes emerging from data collection throughout the research. In the multiple case study approach, there was both within case and cross case analysis (Merriam 1998). Analysis occurred for each case separately with a focus on understanding the case as a whole in order to develop case specific propositions. Data from all methods within each case contributed to the development of propositions for each case, as opposed to data from each method contributing to separate propositions about the case. In this way, feminist validity was enhanced through the process of triangulation. These propositions were then compared and contrasted (or theory tested) across cases. The resulting propositions were also compared and contrasted to propositions generated from Stage Two of the research process.

Data collection occurred prior to analysis, although issues that emerged during data collection also influenced the ongoing research. Therefore, this design reflects both theory generation and testing, and so may be described as both exploratory and explanatory. Acknowledgement that the research adopts neither a stringent inductive or deductive approach is not seen as a drawback of the research but instead is viewed as a more accurate account of research practice, which is often unacknowledged. In addition, it has the advantage of allowing new issues to emerge during the research process as well as propositions to be more fully explored and tested, thus allowing for more rigorous theoretical claims to be made.

Stage One–Informing the Research

This stage included undertaking the literature review, as well as a process of industry immersion and awareness raising of both the research and the importance of the contributions of women in fishing.

While the process of industry immersion and awareness raising about the research continued to occur throughout the research, activities involved in this early research phase included sending an introductory letter to fisheries management agencies and industry organisations; sending an article about the research to key industry magazines; discussing the research with a range of industry people; attending conferences, and visiting several fishing establishments, including aquaculture farms and fishing cooperatives.

This process of immersion: informed the literature review and the research objectives and questions; enabled industry stakeholders, including women in fishing, to have input into the research; facilitated the development of my industry networks; raised awareness of the research and the importance of the contributions of women in fishing; increased my industry familiarisation; and introduced me to the experiences of women in fishing.

Reading for the literature review was directed by the key themes identified in the initial research proposal-the visibility and contributions of women in fishing, the barriers to and opportunities for women's greater participation, and women and industry sustainability. Given the lack of literature on the involvement of Australian women in fishing, my reading included research on the contributions of women in fishing overseas and the contributions of women in farming. Other areas of literature explored in my reading included gender, sustainability, natural resource management and feminist methodology and theory. The literature review was central to the development of the research questions and objectives. The process of reviewing the literature and modifying the literature review continued throughout the course of the research.

Stage Two-Broad Industry Research

Stage Two included a continuation of the process of industry immersion, the collection of statistical data, a review of available documentation on major industry issues and key informant interviews with eight "visible" women in fishing. This Stage was incorporated into the research design to: raise awareness of the research and the importance of the contributions of women in fishing; increase my knowledge of industry operation and issues in order to provide a context for the case study research; compensate for the lack of national information about the contributions of women in fishing; explore the knowledge and views of women in leadership roles in industry organisations; and further develop contacts and networks at a national level to increase the potential for the implementation of concrete strategies.

Data collection included sourcing industry documents and statistical information from fisheries management agencies and industry organisations; and conducting semi structured key informant interviews. Letters requesting information on the representation of women in fishing as owners of licenses, leases and quota; the inclusion of women in fishing in statistics; and policies and practices that recognised women in fishing as clients or stakeholders were sent to state and Commonwealth fisheries management agencies. Letters requesting information on the representation of women in leadership position in industry organisations, membership statistics by gender, and voting and selection processes, were also sent to all national and state peak industry organisations. While responses were received from each of the government fisheries management agencies, even after a reminder letter was sent to peak industry organisations, only two replies were received. A range of industry documents was also collected in this early stage of the research, and documents such as reports on Ecologically Sustainable Development (ESD) and general industry issues, continued to be collected throughout the research process.

Semi structured interviewing enables the collection of specific data while allowing new knowledge to emerge and provides insight into the "...emerging worldview of the respondent" (Merriam 1998, 74). Thus it is particularly suited to both exploratory research and feminist research. Semi structured interviews were conducted with eight women from across Australia in a range of fisheries and positions. Women were selected as key informants on the basis of their activism and leadership which rendered them "visible" in the industry. These interviews with women in leadership positions in industry organisations were included in the research to provide a specific and unique perspective on the contributions of women in fishing. Input from these key informants also contributed significantly to the research direction, research design and outcomes.

A pilot process to improve my skills and ensure the adequacy of questions (Merriam 1998) also occurred during the initial interviews. Suggestions were sought from initial participants on modifications to the interview questions. While these participants felt that no alterations were necessary, after transcribing these interviews, I reflected on the process and conducted a self evaluation. As a result I made some changes to improve my interviewing technique.

Stage Three-Multiple Case Study Design

Stage Three employed a multiple case study design consisting of three cases which explored the contributions of women in the Western Australian Rock Lobster fishery, the Tasmanian oyster fishery, and long lining and trawl fisheries in *Cardea*, the pseudonym used for an Australian east coast fishing village. Given the small size of this village, this pseudonym was adopted to protect the identity of participants. Investigation occurred across the fisheries at both the industry and the enterprise level. Methods employed included semi structured interviews with women in fishing (47), semi structured interviews with representatives from industry organisations and fisheries management agencies (7), focus groups (2 groups, 10 women), observation, and analysis of documents and historical and cultural resources. The inclusion of these secondary sources accounts for quotes and references included in my analysis chapters which are acknowledged as the work of other authors.

The case study approach

The majority of case study methodology texts are based on either positivist (eg Yin 2003a) or interpretive paradigms (eg, Merriam 1998; Stake 1995) and do not incorporate feminist epistemological and methodological principles. Thus, the approach adopted in this research represents a feminist reconceptualisation of the established case study approach. Indeed, this is common practice in feminist research, where the case study method has been employed to suit a researcher's purposes in documenting history and generating theory (Reinharz 1992). As a sociological approach, case studies allow for rich data to be generated through the in depth study of social life (Hakim 2000; Hamel, Dufour and Fortin 1993; Merriam 1998; Reinharz 1992; Stake 1995, 2000). Thus, case studies "retain the holistic and meaningful characteristics of real-life events" (Yin 2003a, 2) and are particularly suited to studying complex phenomenon (Hakim 2000).

A case study may be defined as an end product, a process or research strategy (Yin 2003a), or as a unit of analysis, such as a "bounded system" (Merriam 1998; Stake 1995). Case studies are also inherently flexible (Gillham 2000; Hakim 2000; Stake 1995). They may be based on a single case or multiple cases and are characterised by the use of multiple methods, which are neither necessarily qualitative or quantitative (Yin 2003a, 2003b). Their focus can be individual case histories or community studies; studies of social groups, organisations or institutions; or studies of specific events, roles, relationships and interactions (Hakim 2000). The broad continuum evident in case studies between description, exploration and explanation (Hakim 2000) is conceptualised differently by different authors, particularly in relation to the role of theory (Merriam 1998; Stake 1995; Yin 2003b). Despite this, there is recognition by many authors that case studies are guided by theoretical propositions or questions (Hamel Dufour and Fortin 1993; Merriam 1998; Yin 2003b) and it is this initial use of theory that distinguishes the case study approach from ethnographic and qualitative methods (Yin 2003b).

There are also several common criticisms of the case study approach. Case studies have been criticised as not allowing generalisations to be made (Hamel, Dufour and Fortin 1993; Yin 2003a). This research counters this criticism by rejecting positivist notion of generalisations and adopting bounded generalisation as a form of specificity based on shared social position. Case studies have also been criticised due to the influence of the researcher on the findings (Hakim 2000; Merriam 1998). However, I address this issue by discussing the spilt between subject and object and by employing strategies to ensure feminist validity.

There are several reasons why I chose to adopt the case study approach in this research. Firstly, due to the gynopia and misogyny characterising sociological research,

> "feminist engagement in the production of woman-centred case studies is a greatly needed corrective device" (Reinharz 1992, 168).

Secondly, as a powerful technique for understanding the complexity of social life, the case study approach is particularly useful for adequately incorporating diversity and difference. Thirdly, the case study approach is appropriate where research questions take the form of exploratory questions, where there is a focus on contemporary events and where control is not required over behavioural events (Yin 2003a). Fourthly, the triangulation inherent in the case study approach increases the feminist validity of the research. Employing a multiple case study design also increases the rigour of emerging theory through the process of comparison and contrast between cases, and increases the potential for greater awareness of the contributions of women in fishing over a wider geographical area.

This research employed the case study approach as a flexible research strategy informed by feminist research principles, which incorporated description and explanation and generated theory. The research involved a multiple case study design examining three cases with each case conceptualised as a bounded system (Miles and Huberman 1994, cited in Merriam 1998).

> "Miles and Huberman (1994) think of the case as a 'phenomenon of some sort occurring in a bounded context' (25). They graphically present it as a circle with a heart in the centre. The heart is the focus of the study, while the circle 'defines the edge of the case: what will not be studied'" (Merriam 1998, 27).

For the Western Australian Rock Lobster fishery and the Tasmanian oyster industry, the unit of analysis was the contributions of women in fishing, and the bounded context was the specific fishery determined by species and a particular fisheries management agency. In the third case study of the contributions of women in fishing in long lining and trawl fisheries in Cardea, the unit of analysis remained the contributions of women in fishing. However, the bounded context was not a specific fishery and particular fisheries management agency, but a geographical area.

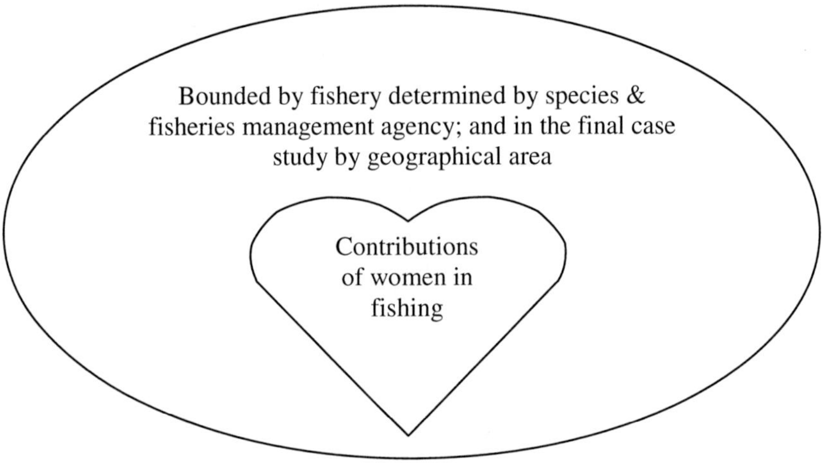

Figure 3-1: Diagram conceptualising case study boundaries

A case study protocol was developed to guide the research process (Yin 2003a), which included a clear description of the unit of analysis, research questions, selection plan for case studies, identification of data sources, outline of data collection activities, arrangements for start up, confidentiality issues, and time and resource estimates.

Selection process for case studies

Case study research focuses on particularization and understanding rather than representation (Stake 1995). Factors to consider in selection therefore extend beyond typical or unusual cases to include "strategic" or "critical" cases (Hakim 2000), diversity and comparison (Rose 1991), individual case characteristics, cases most likely to increase understanding, research purpose, and time and access to fieldwork (Stake 1995). For this research, these principles translated into asking not "What case studies best represent women in the fishing industry?", but instead "What case studies will be most helpful to understanding the contributions of women to the fishing industry?" I then developed a list of relevant characteristics by which to compare different fisheries. These characteristics included: management agency (Commonwealth or State), geographical location, sector (wild catch or aquaculture), size of fishery (number of licenses, leases, boats), value of fishery, age of fishery, demographic diversity, management controls (input or output), stock status, current fishery issues relating to sustainability, and the work organisation involved in harvesting (land or sea based).

Ideally, the selection process would have begun with a list of the characteristics of each of the approximately 150 State and Commonwealth commercial fisheries. However, there is no available centralised information about fisheries. In addition, fisheries are in a constant state of flux-fisheries close, new fisheries emerge and responsibility for managing a fishery may shift from between State and Commonwealth fisheries management agencies. I then considered which fisheries would best help me to understand the contributions of women in fishing in the context of fieldwork opportunities and practicalities. Due to the difficulties in conducting relatively short periods of fieldwork at a considerable distance from my local area and accessing women in fishing, it was essential to have the support of key contacts in each case study.

Early in the research process, my attendance at a conference generated interest from several Tasmanian women oyster farmers who were keen to participate in the research, and I also developed relationships with and had support from several oyster industry organisational and fisheries management agency stakeholders. I therefore decided to include the oyster industry in Tasmania as a case study site. At this conference I also received strong support from WINSC and key industry stakeholders to conduct a case study in the Western Australia (WA) Rock Lobster fishery.

After conducting these two case studies it became apparent that, despite the history of diverse ethnic involvement in the WA Rock Lobster fishery, women from a variety of ethnic backgrounds were not well represented amongst participants. Therefore, to include women who were less "visible" and not as likely to become involved through either direct advertising or purposive snowball sampling, and reflecting flexibility in altering the research design (Stake 1995), I decided upon a third case study.

The third case study, focusing on the contributions of women in long lining and trawl fisheries in Cardea, was selected on several grounds.

Fishing in this village was established by the Italian community in the 1920's and remains dominated by Italian families. This ensured that ethnic diversity was incorporated into the research. As this village was close to my residential location, time was not a constraint for fieldwork allowing a level of trust to be developed amongst potential participants and ensuring wider participation of less visible industry women.

Table 3-1 outlines some of the divergent characteristics between the three case studies where research was undertaken.

Table 3-1: Characteristics of the three Case Study fisheries

Characteristic	WA Rock Lobster fishery	Tasmania oyster fishery	Cardea fisheries
Fisheries management agency Commonwealth-Australian Fisheries Management Authority-or State	State (Western Australian Fisheries)	State (Department of Primary Industries, Water and Environment)	Commonwealth (3 fisheries)– Commonwealth Trawl Sector, Southern Bluefin Tuna longline fishery, Eastern Tuna and Billfish Fishery
Age of fishery	Established in the 1890's	Established in the 1950/60's & expanded in the 1990's	Established in the 1920s-1930s
Geographical location	Western Australia	Tasmania	Small village east coast Aust.
Sector (wild catch or aquaculture)	Wild catch	Aquaculture	Wild catch
Size of fishery (number of licenses, leases, boats)	Approx 600 Managed Fisheries Licenses	Approx 90 leases	Approx 15 boats at the wharf
Demographic diversity	History of diverse ethnic involvement	No history of diverse ethnic involvement	Italian fishing community
Harvesting	Sea based with baited pots	Estuary/ bay based on racks	Trawling & long lining

Other features & industry issues at time of research	First fishery internationally to receive Marine Stewardship Certification; longest running management plan of any Aust. fishery; undergoing management review under National Competition Policy requirements	Implementing Environmental Management Systems (EMS); amount & quality of water from up stream estuary users	All Commonwealth fisheries subject to restructuring

Western Australian Rock Lobster fishery

The Western Australian Rock Lobster fishery is a sea based, seasonal fishery that targets western rock lobster (Panulirus Cygnus) using baited pots from boats (WA Fisheries 2002) which, at the time of the research, was operating from 15th November through to 30th June. Established in the 1890's, there was a range of cultural and historical resources on the fishery. The fishery extended from North West Cape in the north, to Cape Leeuwin in the south, and was divided into three fishing zones (WA Fisheries 2002). The fishery is Australia's most valuable single species fishery with major export markets, with lobsters sold to registered processors who market and export the harvest.

The fishery is managed by WA Fisheries and, at the time of the research, there were approximately 600 Managed Fisheries Licenses in the fishery. In 1963 the fishery was declared a limited entry fishery and as such, has the longest running management plan in Australia of all commercial fisheries. At the time of the research, the stock status was described as "fully exploited within the bounds of sustainability" (WA Fisheries 2002). In 2002, the Western Australian Rock Lobster fishery was the first fishery internationally to receive certification as sustainable from the Marine Stewardship Council (MSC), the world's only independent international fishery certification and eco-labelling program that is consistent with Food and Agriculture Organisation (FAO) guidelines (MSC 2006).

At the time of the research, the fishery had several management controls including restrictions on the total number of commercial rock lobster pots able to be used in each zone, size limits and the protection of breeding females. A Managed Fishery License, which must have at least 63 pots attached to it, was required to operate in the fishery. While the limited number of total pots in the fishery provides restriction, there is significant variety in how the pots attached to a license may be comprised-pots may be leased or owned by the license holder, license owners may own a proportion of the pots and lease the remainder, and several investors may each own a small number of pots associated with a license. The owner may operate the license, the license may be leased out to an operator, or an owner may employ a skipper.

A review of management controls had commenced in 2002 under National Competition Policy (NCP) requirements and was due to be completed in 2007. This review had already seen the lifting of the maximum 150 pot rule for each Managed Fisheries License in 2003. Other areas under review at the time of the fieldwork were lifting the minimum 63 pot rule and changing to a quota, or output management control system. At this time, these proposals generated significant concern in the industry. After fieldwork was completed in 2003 the minimum 63 pot rule was lifted. A decision about going to quota management was expected to be made in 2007 by the state Minister responsible for fisheries.

At the time of the research, there were several industry organisations associated with this fishery. These included the Western Australian Fishing Industry Council (WAFIC) responsible for general industry issues, such as occupational health and safety; the Western Australian Rock Lobster Council (RLC) comprised of the Presidents of each of the 14 local Professional Fishermen's Associations; and the Rock Lobster Industry Advisory Council (RLIAC), the Ministerial advisory council that included representatives from government and industry.

Tasmanian oyster fishery

The Tasmanian oyster fishery was established between the 1950s and 1960s and experienced significant expansion in the 1980s and 1990s. Growers purchase "single seed" oysters from hatcheries and grow them to size in estuarine waters. The fishery produces about 2.6 million Pacific oysters (Crassostrea gigas) from a total area of 1500 hectares and currently generates twenty million dollars annually.

At the time of the research, Tasmania had approximately 90 oyster lease holders. The fishery was managed by the Tasmanian Department of Primary Industries, Water and Environment (DPIWE) and regulated under the Living Marine Resources Management Act 1995 and the Marine Farming Planning Act 1995. Under this later Act, marine farm planning is a process which has identified all waters suitable and available for aquaculture.

Issues in the fishery at that time included the introduction of Environmental Management Systems (EMS) with a specific focus on water quality issues. EMS templates translate the principles of ESD into operational practices, which includes components such as hazard analysis and risk management. Just prior to fieldwork in 2004, many east coast growers had been affected by excessive freshwater flooding and extensive stock losses meant some operators were forced to leave the fishery.

Industry organisations relevant to this fishery included the TFIC, of which all lease holders are members; the Tasmanian Shellfish Executive Council, which was a member of the Tasmanian Agriculture Council; and several smaller local industry organisations such as the Tasmanian Marine Farmers Association and Bruny Island Shellfish Association.

Long lining and trawling in Cardea

Originally a timber and agricultural area, fishing prospered in Cardea in the 1920's with an influx of Italian immigrants, who settled in the town due to the abundance of fish. The families who fish today are descendants of these original Italian fishing families. In I971, there were 74 Italian fishing families each with their own boats. However, at the time of the research, there were only 15 remaining boats in the port. Initially, fishermen concentrated on the trawl fishery but later established tuna fishing in the area. Over time, the trawl fishery, which targets multiple species using nets, has become increasingly economically unviable and undergone several restructures. Many families have moved into other ventures. Originally, tuna were caught in large numbers and sent to the canneries. However, in subsequent years Blue Fin Tuna has become particularly lucrative and quota was allocated in the 1980's. Blue Fin Tuna, now caught on "long lines", is exported to Japan for the sashimi market. Other species caught by long lining include yellow fin and striped tuna.

Both the trawl and long line fisheries operate all year and are managed primarily by the Commonwealth fisheries agency, AFMA, although NSW Fisheries has a small jurisdiction over waters 4 nautical miles from the shore. These fisheries are overseen by Management Advisory Committees (MAC's) comprised of members from a range of sectors. The Cardea Fisherman's Cooperative was established in the 1950's by the Italian fishermen and still markets and transports fish for members. While tuna is mainly exported, the trawl fishery supplies Melbourne and Sydney fish markets. Both the long line and trawl fisheries have been involved in the process of negotiating compensation for displaced fishing effort as a result of the creation of the Marine Protected Areas in the south east region (AFMA 2005). This compensation was included in the release in late 2005 of a $220 million dollar adjustment package by AFMA, which also, through a process of voluntary tendering, aimed to "encourage" individual fishing businesses to exit from all Commonwealth managed fisheries (AFMA 2005).

Case Study Log

A case study log which recorded the resources collected in each case study was developed (Yin 2003a). Details are recorded in Table 3-2.

Accessing participants

Four semi structured interviews were held with representatives from WA Rock Lobster fishery fisheries management agencies and industry organisations using an interview schedule. In terms of accessing women in this fishery, I utilised a publicly available list of the 600 Managed Fisheries License holders. However, these license holders were listed as business entities with no personal details of the proprietors. In response to a mail out to those on this list seeking women who were interested in participating and forty seven flyers were returned representing a 9.4% response rate. However, this response rate may be inaccurate as there is no data indicating the number of family fishing businesses as opposed to corporate entities operating in the fishery. One further factor that may have contributed to this low response rate was the lack of confidence of, and recognition by, women and/ or their partners about the value of their contributions. In addition, some women in fishing may not have received the information as, in most instances, the envelope was addressed to a business. Many women who actually participated in the research explained that they had discussed the research with other women in fishing as a result of the letter. Several women also were involved as a result of contacting me after reading about the research in industry magazines, or after a suggestion was made that they participated by another industry contact known to them.

A multi faceted approach similar to that employed in the Western Australian Rock Lobster fishery was adopted to access potential participants in the Tasmanian oyster industry. Information explaining the research and a flyer was sent to the 90 lease holding entities. As a list of lease holders was not publicly available, this information was sent out through TFIC. However, only one response was received. One contributing factor in this case study may be the geographical isolation between women, which reduced the potential for discussion about the research. I then proceeded to employ substantive purposive snowball sampling through WINSC, other industry organisations and women who had already expressed interest.

The contributions of women in Cardea was a case study designed to target women in fishing unlikely to respond to direct advertising, and so, information was not sent to all license holders in Cardea. Initially I had discussions with a prominent local community leader, who had connections with the industry and who passed on the names of two women. I contacted and arranged interviews with these women, each of whom passed on more names. A high proportion of women I contacted through this ongoing process (five) declined to participate. I felt this reluctance of women in Cardea to participate may at least be partially attributable to my being an "outsider" in the Italian community.

I also placed an article in the local paper and although three participants commented that they had read it, I received no responses. Interviews were held with two representatives from the Cardea Fishermen's Cooperative. I also requested to attend the monthly meeting of the Cooperative, but was informed that these meetings were "confidential" and only members were able to attend. No formal interviews were held with representatives from fisheries management agencies or industry organisations as the purpose of the case study which was focused on exploring micro social relations specifically pertaining to ethnicity within the bounds of a small geographical area. However, informal discussions were held with fisheries management agency representatives and a MAC representative to provide information on the trawl and long lining fisheries.

Participants

Thirty four women from the Western Australian Rock Lobster fishery participated in the research. Seventeen women from the Tasmanian oyster fishery participated in the research. In contrast to Western Australia where coastal villages and fishing communities had developed as a result of the rock lobster fishery, women involved in the Tasmanian oyster fishery were more widely dispersed geographically.

Six women from the Cardea participated in the research. Two of the women interviewed were involved in the long line fishery, one woman was involved in the trawl fishery and three women were involved in businesses which held licenses in both long lining and trawl fisheries. A notable feature relating to participants in this site was the extent and longevity of extended family business involvement and family connections. As a result, all partners had a lengthy involvement with the fishing industry. All six of these women's partners commenced fishing with relatives as teenagers; four were 2nd generation fishers; and 2 were 3rd generation fishers.

Table 3-3 provides demographic details of women who participated in the research. Where the figures presented in the tabular representation do not add to the total number of women in fishing involved in each case study, women either chose not to provide the information, or time constraints meant that the topic was not prioritised and so not covered in interviews.

Table 3-2: Case study data log summary

Data collection	WA Rock Lobster fishery	Tasmania oyster fishery	Cardea fisheries
Semi structured interviews with women in fishing	24	17	6
Focus groups	2 groups x 5 women in fishing		
Semi structured interviews with representatives	4 from WA Fisheries, RLIAC, WRLC and a Professional Fisherman's Association	3 with fishery management agency and industry organisation representatives from DPIWE, TFIC, Tasmanian Shellfish Council, Tasmanian Agriculture Council and Tasmanian Marine Farmers	2 Cardea Fishermen's Cooperative representatives Informal discussions with officers from fisheries management agencies Informal discussions with a MAC representative
Observation	Attendance at the 2003 Rock Lobster Congress	2 oyster farm tours and a fishery networking dinner Informal discussions with women in fishing	Attendance at an AFMA public meeting held to discuss proposed fisheries management changes
Industry documents	Range	Range	Range
Historical and cultural resources including:	*The Western Rock Lobster Fishery: A history of the fishery* (Gray 1999) *Cray Tales* (de Monchaux 1999) *Abrolhos Islands Conversations* (France, Mitchell and Wright 1998) Photographs/ newspaper articles from local history collections Fishing magazines		Local newspaper articles 2 videos on fishing in Cardea Sources from the Cardea Library local history collection

Table 3-3: Demographic characteristics of women in fishing participants

n=number of women *m=median r=range*	**Rock Lobster fishery women** n = 34	**Tas. oyster fishery women** n = 17	**Cardea women** n = 6
Age	r = 27-64 yrs/ m = 45 yrs	r = 25-65 yrs/ m= 45.5 yrs	r = 26-61 yrs/ m = 48 yrs
Highest education level			
University	n = 7	n = 6	-
Vocational-TAFE	n = 5	-	n = 2
6 yrs High School	n = 8	n = 2	n = 2
5 yrs High School	-	n = 2	-
4 yrs High School	n = 11	n = 4	n = 2
Ethnic background			
Australian	n = 17	n = 14	n = 4
Croatian-Australian	n = 3	-	-
European	n = 2	-	-
Italian	-	-	n = 2
Italian-Australian	n = 2	-	-
Latvian-Australian	n = 1	-	-
Peruvian	-	-	n = 1
Partner's ethnic background			
Australian	n = 17	n = 13	-
Italian	n = 3	-	n = 6**
European	n = 1	n = 1	-
Slavic	n = 1	-	-
Aboriginal/ Malaysian	n = 1	-	-

Marital status			
Married	n = 26	n = 13	n = 6
Second marriage	n = 2	n = 1	-
Widowed (sole operators)	n = 2	-	-
Divorced/separated (still in fishery)	n = 1	n = 2	-
De facto	n = 1	n = 1	-
Single	n = 1	-	-
Children			
Women with children	n = 30 n of child. = 70	n = 16/ n of child. = 47 r of child.’s age = 1-40yrs	n = 6 r of child.’s age = 1-41yrs
Women with no children	n = 2	n = 1	-
Women who were pregnant	n = 3	-	-
Domicile			
Capital city vicinity	n = 11 (Perth)	n = 5 (Hobart)	-
Coastal villages/ rural coastal areas	n = 6	n = 4	n = 6#
Small-larger coastal towns	n = 17	n = 4	-
Duration of involvement in fishery	r = 2-53 yrs/ m = 25 yrs	r = 3- 40 yrs/ m = 12.5 yrs	r = 3- 43 yrs
Entry into fishery			
Entry through marriage	n = 27	n = 7	n = 5
Bought into oyster business with partner	-	n = 9	-
Born into fishing families/ married fishers	n = 5	n = 1	n = 1
Born into fishing family/remained in business	n = 1	-	-
Deckhand prior to marrying a fisherman	n = 1	-	-
Status in the business		n of businesses=19	n of businesses=5
Legal ownership in business &/or fishing capital with partner &/or other investors	n = 29	n = 11	n = 3

No legal ownership in business and/or fishing capital	n = 4	n = 3	n = 3
Involved in family trusts with children	n = 3	n = 3	n = 2
Sole owner operators***	n = 4	n = 2	-
Contributes in a manual capacity- paid employee	-	n = 2	-
Extended family involvement in fishery			
Extended family involved	n = 11	n = 1	n = 6
Sons involved in fishery	n = 9	n = 3	n = 3*^
No extended family involved	n = 9****	n = 11	-
Daughters involved in the business	-	n = 1	n = 1**^^
Paid work outside the fishery*			
Self employed-non fishing business	n =2		
Employed part time work*^^*	n =3	n =4	n =3*^*
Employed full time work	n =2	n =4	
No paid employment outside fishery	n =25	n =6	n =3^

* There were large income differentials between participants, particularly between families who owned and operated licenses and families who leased licences and pots. Women who leased licenses were working outside their business to support their households. **Five having one or both parents born in Italy. ***Women with deceased partners; one woman who inherited fishing license from her father; and single women in partnership with other investors. ****Some of these women had younger children whom they felt may go into the fishery. *^ 4th generation fishers. **^ This woman had no sons-eldest daughter married a fisherman who was skippering a boat involved in the family business. *^^* These women had school age children. *^* As a secretary and hairdresser working to support households with combined incomes of $30 000 and $45 000, and both with young children. ^ Not worked outside the fisheries since they were married, and one woman who worked for 30 years as a pharmacy assistant retiring when business debt free. # Three women born locally, two moved into area with their families as children, and one relocated to be with partner.

Methods employed in case studies

There were a range of methods included in the case study research. These included semi structured interviews and focus groups with women in fishing; semi structured interviews with fisheries management agency and industry organisation representatives; observation; and sourcing industry documents and cultural and historical resources. The use of several methods ensured triangulation which addresses feminist validity. Each method was employed with a focus on exploring the contributions of women in fishing.

The period of data collection varied between case studies. Fieldwork in the Western Australian Rock Lobster fishery and the Tasmanian oyster fishery occurred over a short intensive period of three weeks and ten days respectively. Fieldwork in Cardea occurred over a period of several months. However, sourcing documents and historical and cultural resources for these case studies continued throughout the duration of the research.

Semi structured interviewing

Semi structured interviews were governed by the principles outlined previously in this chapter. To ensure that I entered the field with some knowledge about the fisheries, in addition to reading on each fishery, interviews with fisheries management agency and industry organisation representatives were held prior to fieldwork commencing.

Women in fishing who participated in semi structured interviews were given the choice of where they would like to be interviewed and while the majority of interviews were conducted in women's homes, several were conducted in cafes and workplaces. Interviews ranged in length from 45 minutes to two hours, were tape recorded and transcribed. All women received an information sheet and signed a consent form. To ensure that the research and participation was understood, I explained the research and went through the forms with each woman. In addition, information on a telephone counselling service was also provided. Fisheries management agency and industry organisation representatives also signed a consent form prior to interviews being conducted.

Initial interviews served as pilot interviews and several modifications were made after feedback from women involved in these interviews. I also had asked these women to complete and comment on a form which was designed to provide financial and business information. This was done on the basis that the women involved may feel less apprehensive about revealing this information on paper rather than in response to direct questions. However, the feedback from women was that they were not able to complete it accurately as the sheet did not cover the complexity of their financial arrangements. In addition, several participants chose not to complete it. After these initial interviews, I decided to abandon the financial questionnaire sheet.

Industry documents/ historical and cultural resources

Industry documents and information were collected from management agencies, industry associations and were accessed through the internet. In two case studies, newspaper articles and photographs were located in local library collections. Books, videos and information on the cultural and historical development of the fishery were sourced from libraries, the internet and women who were involved in the research. Resources reflecting the historical and cultural development of the fishery were not available in the Tasmanian oyster fishery case study. This can be attributed to the geographical dispersion of oyster farms across the State and the relatively recent nature of the industry.

Observation

Observation provides a first hand account of behaviour and includes attention to the physical setting, participants, activities, conversation, subtle factors and my own behaviour (Merriam 1998). As an observer, I ensured that my role was clear to participants. However, my position fluctuated on the participant/ observer continuum depending on levels of familiarity and the sensitivity of the interactions I observed and in which I participated (Merriam 1998). Notes were taken during periods of observations and later written up comprehensively into my research journal. These notes included a description of settings, people and activities, the nature of what was said and my comments and feelings (Merriam 1998).

Focus groups

The central feature of focus groups is the interaction of participants, which provides insight into how meanings are expressed, shared, modified, constructed and contested (Kitzinger and Barbour 1999; Quible 1998; Wilkinson 1999). Focus groups may then be seen as a particularly useful strategy for feminist research which makes "what is invisible to many women visible" (Pini 2002b, 339), and where individual and collective experience generates discussion from which subjectivities may alter (Pini 2002b; Wilkinson 1999). Therefore, focus groups can facilitate political intent and action (Baker and Hinton 1999; Kitzinger and Barbour 1999; Pini 2002b). This process was reflected in one focus group where, towards the end of discussion, women were asking themselves, and each other, why they had not been more involved in industry decision making and expressed their intention to become more involved.

Only two focus groups were incorporated into the case study research, both of which were with women in the Western Australian Rock Lobster fishery. Unfortunately, including a greater number of focus groups in the research was not possible due to the difficulties in accessing women in fishing, their geographical dispersion, and in Cardea, issues relating to a small group of related women involved in ethnic fishing families. One of the focus groups involved two mothers-in-law and daughters-in-law.

Issues relating to positioning and power are relevant in the use of focus groups (Chiu and Knight 1999; Kitzinger and Barbour 1999). While focus groups may reduce the power hierarchy between the researcher and participants (Morgan and Krueger 1993; Wilkinson 1999), familiarity raises issues of power and positioning *amongst* participants. Familiarity may affect opportunities for empowerment (Pini 2002b) and may either facilitate or hinder open discussion (Kitzinger and Barbour 1999). Drawing from my research journal, my impression was that the group where participants did not know each other facilitated a greater opportunity for altering subjectivities. In the group where participants were connected through marriage and friendship, one daughter-in-law was not as free to participate as she may have been had her mother-in-law not been involved. I felt that this mother-in-law exercised power by censuring her daughter-in-law's level and type of participation. Adopting a flexible approach to group facilitation (Kitzinger and Barbour 1999) and, despite employing my skills in group facilitation and attempting to directly engage the daughter-in-law, I experienced the mother-in-law's capacity to exercise power in this context as greater than mine as the researcher.

Ethical Issues/ Methodological Difficulties

Ethics approval was granted for this research. Corresponding with approval, participants involved in interviews and focus groups were provided with information sheets and signed consent forms prior to participating. Women, who engaged in interviews and focus groups, were also provided with details of a telephone counselling service. The voluntary nature of participating, including the tape recording of interviews, was discussed.

The importance of maintaining confidentiality was highlighted to all participants. Prior to the commencement of focus groups, women were asked for their commitment in this respect. In the data collected, where identifying features emerged such as names of individuals, businesses or boats, or any situations which could identify any person, details were altered in the research write up. A pseudonym was employed for the east coast fishing village to protect the identity of women involved in the research given the established profile of this fishing village, its small size, and the small number of interrelated families in the fishing industry in the village.

The methodological difficulties experienced in the course of this research have been raised throughout this chapter. In sum, these difficulties include accessing and recruiting women in fishing, women's busy personal schedules, women's reluctance to discuss financial matters, a lack of information available from fisheries management agencies and industry organisations, the lack of responses from industry organisations, the lack of historical and cultural resources in the Tasmanian oyster industry, issues of power and positioning, fieldwork constraints, interviews being held in locations such as cafes and workplaces which were not conducive to transcription, and ethical dilemmas about what to include and what not to include as research data.

Data Analysis

Demographic information which emerged from interviews was recorded on a spreadsheet. The analysis of all other data was undertaken utilising NVivo software and according to the principles of qualitative analysis and coding (Merriam 1998). Themes and categories were constructed to reflect the purpose of the research and were exhaustive, mutually exclusive, sensitising and conceptually congruent (Merriam

1998). These categories were formed using the constant comparative method, where text is examined according to similarities and differences in a process of comparison (Merriam 1998). Care was taken to ensure that the number of categories were not too few, signifying too high a level of abstraction, or too many, signifying analysis as embedded in too concrete a description (Merriam 1998).

Analysis began with a preliminary coding of broad themes relating to the research questions. These themes included the visibility of women in fishing, their contributions (past, present, future) and involvement in decision making (past, present, future), barriers to women’s contribution (past, present, future) opportunities for women’s contributions (past, present, future), and women in fishing and sustainability. Secondary coding was then conducted which reflected the critical constructs inherent in the theoretical position adopted in this research including: power, knowledge, resistance, discourse, subject positions and subjectivities. This coding was undertaken with a commitment to reinscribing women with agency, to not take things for granted, and to changing the existing state of things (Jupp and Norris 1993).

The post structural understanding that subjectivities are negotiated in and through discourse, and that meaning is constructed through language, was evident in the process of analysis. Data was transcribed with as little “cleaning up” of the script as possible (Myers and Macnaghten 1999) and analysis conducted with a focus on interaction, language, the construction of meanings, and how they were contested, expressed and modified (Green and Hart 1999; Kitzinger and Barbour 1999; Myers and Macnaghten 1999; Pini 2002b; Quible 1998; Wilkinson 1998b; 1999).

To assist practically with the theoretical analysis of data, I applied the following questions influenced by Foucauldian thought and suggested by Jupp and Norris (1993).

> “1. What public and/or institutional discourses are important in terms of knowledge of what is ‘right’ and what is ‘wrong’?
>
> 2. In what kinds of documents and texts do such discourses appear?
>
> 3. Who writes or speaks these discourses and whom do they represent or purport to represent?
>
> 4. What is the intended audience of such discourses?

> 5. What does a critical reading of these documents uncover in terms of:
> a) what is defined as 'right' and 'wrong' and therefore what is seen as problematic;
> b) what therefore is offered for what is seen as problematic;
> c) what therefore is seen as the solution?
>
> 6. What does a critical reading of these documents tell us about
> a) what is not seen as problematic:
> b) which explanations are rejected or omitted;
> c) which solutions are not preferred?
>
> 7. What alternative discourses exist?
>
> 8. How do these relate to 'internal differentiation' within and between semi-autonomous realms of control?
>
> 9. What does a critical reading of these alternative discourses tell us?
>
> 10. Is there evidence of negotiation with, or resistance to, dominant discourses?
>
> 11. What is the relationship between the discourses and social conflict, social struggle, hierarchies of credibility, order and control, and most of all, the exercise of power?" (Jupp and Norris 1993, 50)

Jupp and Norris (1993) also suggest the following question:

> "12. Are discourses, knowledge and power pervasive or reducible to class, class conflict and struggles refracted through one source, the state? (Jupp and Norris 1993, 50)"

However, this was modified into the following question to incorporate women's agency and to reflect gender as an organising principle of society.

> 12. How pervasive are particular discourses, knowledge and power, and how are women positioned in these discourses?

Again power and positioning emerge as vital considerations in the research process through the final question posed by these authors.

> "Does a critical reading of dominant discourses inevitably lead to the construction of alternate and competing discourses by the social researcher?" (Jupp and Norris 1993, 51)

Answers to the research questions emerged from the data and through theory building (the linking of propositions) (Merriam 1998). Theory verification occurred by linking categories together (Merriam 1998). Analysis attempted to constantly compare propositions emerging from the themes until the propositions explained all known cases of the phenomenon (Merriam 1998). Analysis and theory building also occurred with attention given to the observation of Wilkinson (1998a) who notes the tendency for analysis to demonstrate "slippage" between the essentialist and social constructionist frameworks.

Conclusion

This chapter has explained the research process and provided an overview and discussion on the feminist methodological principles adopted in this research. In concluding, I reaffirm my commitment to feminism as a form of politics which prioritises the voice of women.

> "Feminism is premised on a particular sort of effort of attention to the experience of other women, which is why it does matter 'who is speaking'. This effort of attention to what we share and to the ways in which we differ, brings the speaker, her world, her knowledge and her contemporary or historical silences into connection with the theorist's experience and knowledge production. It is in part because feminism is concerned not just with 'difference' but with the variable moral and political weight, and the different dynamics, of difference, that it both exceeds the terms of modernist thinking and resists absorption into the discourses of postmodernity" (Ransom 1993, 144)

In the following four chapters, applying the feminist methodological and theoretical principles outlined in this and the preceding chapter, I present my findings relating to the contributions of women in family based businesses in the Australian commercial fishing industry.

CHAPTER FOUR

A MASCULINE INDUSTRY DISCOURSE

Introduction

This chapter explores the nature of gendered power relations in the dominant industry discourse in the Australian commercial fishing industry and how women are positioned within this discourse. In doing so, it addresses the visibility of women in fishing and identifies barriers to their greater industry participation. Previous authors have recognised the gendered language and occupational privileging in fishing (Dixon, Lowery and Sabella 1984; Fink 1990; Lambeth et al 2002; Yodanis 2000) and farming (Alston 1995b, 2000; Hogan 1994); and women's marginalisation in family businesses in farming and fishing based on patrilineal inheritance practices (Nadel-Klein 1988; Shortall 1999; Thompson, Wailey and Lummis 1983).

The first section in this chapter explores language as a discursive site (Weedon 1989; Weiner 1993) to deconstruct the apparently apolitical nature of the construction of "industry", and as a means of revealing the gendered industry power relations and the implications of this language for the positioning of women in fishing. I then explore the gendered power relations in women's legal ownership status of businesses and capital, the practice of patrilineal inheritance in the fishing industry and the implications of this practice for women (Shortall 1999). This is followed by a more thorough exploration of the subject positions of fisherman and fisherman's wife, and an exploration of the normalising discursive regulatory mechanisms that exclude women in fishing from occupational privileging.

This exploration draws on the premise of standpoint theorists that women's experience is shaped by everyday life and that the nature of exclusion is real (Hartsock 2004a, 2004b; Ruddick 2004; Smith 2004a, 2004b). In addition, I draw from the premises of post structural theorists that Western discourse is characterised by binaries which are hierarchically and relationally constructed (Beasley 1999), the productive nature of power (Foucault 1980) and the recognition that there are internal sources of empowerment (Deveaux 1994).

Language as a Discursive Site

The fisherman and his wife

Language is a powerful discursive site, with the terms fisherman and fishermen clearly gendered and privileging men with occupational status. These terms were used prolifically throughout the research, both in industry documents and by participants, to describe both men who harvest fish and industry stakeholders. While textual sites are increasingly substituting the term fisher/s, often the textual site often remains gender blind by referring primarily or exclusively to the act of harvesting (eg, Jusseit and Robinson 2003). Furthermore, there are significant textual slip-ups in fisheries management documents. For example, "fishermen" are named as stakeholders (eg, Fletcher et al 2005, 82), and, in a social impact assessment in the Rock Lobster fishery, women in fishing are referred to as "fishermen's housewives" (Huddleston and Dew 2003, 86).

Some women who harvest fish as an occupation, also identify themselves as "fishermen". An industry magazine describes the position of one of these women, who had inherited her father's fishing license and had fished for 30 years.

> "Political correctness is also a **triviality in which she has no interest. In her language, it doesn't exist. She refuses to acknowledge the terms 'fisherwoman' or 'fisherperson',** or even plain 'fisher'. She **asserts that gender didn't apply to her work, which was as a fisherman**...." (Leach and Twycross 2003, 35).

This ardent claim to "fisherman" status may be interpreted as an indication of the struggle by these women to be acknowledged *as equal* by men who fish and as resistance to being re/constructed by these men as lesser based on gender, and as a *"woman fisherman"*.

"Fisherman" is the subject position almost exclusively occupied by men in the fishing industry, and "fisherman's wife" is a subject position occupied by the majority of women who are involved in family fishing businesses and who are in a relationship, including marriage, with these men. These men are privileged with occupational status, while women are identified by their marital status and are positioned as secondary in the industry. As, such the category "industry" is inherently gendered. The pervasiveness of women's secondary industry status was also reflected and perpetuated in language used by the majority of women in fishing, including one woman in industry decision making, who identified herself and other women in fishing as "fishermen's wives" and "just a housewife".

Furthermore, as in the agricultural industry where farming includes more than physical labour (Alston 1995b; Fink 1990), the activities involved in maintaining a family fishing business involve considerably more than harvesting fish. These activities are undertaken by women in family fishing businesses, but are typically absent in common understandings about what comprises the "fishing industry".

> "I definitely think the women play a huge part in **not necessarily the industry**, cause … **I classify the industry as…harvesting** the crayfish out there…" (woman, rock lobster fishery, license owner, in her thirties).

This woman inherited and solely owned her rock lobster license. She managed her business and her husband harvested lobsters. Despite the potential for this woman to resist a secondary positioning, she commented that being a "successful fisherman" is contingent on having a "decent wife". Deconstructing this language reveals a regulatory discursive function, where women who collude and undertake the activities discursively assigned to the subject position of "fisherman's wife" are rewarded through the construction of being good or "decent" wives.

"Behind the scenes"

Women's subjectivities are constituted in, and by, the language they use. While women in fishing often acknowledged the extensive contributions made by themselves and other women in fishing, their secondary status was revealed in language positioning themselves as being "behind the scenes", "behind the fishermen" or "in the background". However, language is also a site of discursive resistance. One woman, who entered the industry in her fifties in a joint venture with her second husband, corrected her use of language to reflect her resistance to the secondary positioning of women in the fishing industry.

> "Where would the fellows be if there wasn't a woman behind them really? Well **not even behind them, I shouldn't say that.** If there wasn't a

> **woman with them....A woman beside them**, but it's **always seen as being behind"** (woman, oyster fishery, in partnership, second marriage, in her fifties).

Numerous references were also made to women in fishing not being heard (having voice) or seen (being visible).

> "They are sort of the backbone, but the **silent** backbone" (rock lobster woman, in her thirties).

> "...we really contribute towards the industry...**even if our faces are not there.** Yeah, we are in the background working" (woman, oyster fishery, in partnership, ethnic background, in her forties).

Despite the essential contributions made by women in fishing to their businesses, industry, families and communities, this lack of visibility reflects and perpetuates the secondary status of women in fishing and the lack of recognition for their contributions.

> "...there wouldn't be a lot of the fishing industry enterprises at all if it wasn't for the women in the background, but **they don't get much credit for that"** (woman in fishing, key informant).

Textual and visual sites

Masculine industry dominance is often reflected in and perpetuated through visual images and text in the media, where typically, fishing industry stakeholders are represented as the collective of "fishermen" with no reference made to women. Local media coverage of proposed contentious management changes in Cardea reflects this masculine appropriation of fishing at both a textual and visual level. Visual images show men who harvest fish at either the wharf or on their boats-the two main sites of masculine industry power in the wild catch sector-and text refers only to "fishermen". Visual representations of women in fishing often reflect the gendered power relations embodied in the dominant masculine industry discourse and women in fishing tend to be in the home. A 1970s television documentary made of an Italian fishing family from Cardea was dominated by images of the "fisherman" on the boat, at sea and on the wharf. In contrast, his partner, who spoke about her extensive involvement in the industry, is interviewed while cooking in the kitchen. The pervasiveness of gendered power relations in the dominant masculine industry discourse is demonstrated in the follow up documentary on the same family, filmed in the early 2000s, where again, this woman is interviewed while cooking in their kitchen.

In some instances, visual images in the fishing industry depict women generally as sexual objects. One example is found in the Australian Aquaculture Yearbook (NAC 2002), a technical document which includes advertisements targeted to fishing industry operators. This document features images of men engaged in practical aquaculture activities. There are no images of women involved in practical aquaculture activities or in business management. The back cover features a full page advertisement for the Sydney Fish Markets featuring a woman in a bikini (Figure 1-1), with the suggestive headline "Show us your fish". This sexual objectification of women in fishing is historically pervasive, as can be seen from the front of an earlier double sided advertisement (Figure 1-2) which was inserted into a 1976 industry magazine (GEC Diesels Australia Ltd 1976), targeted at fishing industry operators. This demonstrates the targeting of advertisements primarily to men in fishing, despite the fact that many women involved in this research were responsible for some aspects of business management and marketing and selling their fish. Furthermore, while the earlier advertisement is targeted to the historically male dominated wild catch sector, this sexual construction of women has emerged in the more recently established aquaculture sector.

In the last decade a small number of media profiles on women in fishing have been published in newspapers and magazines. However, this profiling tends to focus on the general contributions women make to their family fishing businesses, and typically does not include women's activism or views on significant industry issues. While this profiling increases awareness of the contributions of women in fishing, the exclusion of women's views on industry issues perpetuates their positioning as secondary and supporting, and reinforces their lack of stakeholder status. Women in paid employment in technical and scientific positions in the fishing industry are also becoming more prevalent in visual industry representations. However, women working in these capacities may occupy this discursive space, as they do not have the same potential as "fishermen's wives" to usurp the masculine occupational privileging in the industry by appropriating the practice of fishing. Text and visual images as discursive sites reflect and perpetuate the invisibility of women in family fishing businesses and the subsequent assumption that they are not involved in the industry.

"... they [women] really **don't play a big part in the industry** down here" (male representative, Cardea Fishermen's Cooperative).

Figure 1-1: *Sydney Fish Market* advertisement featured on back page of the *Australian Aquaculture Yearbook* (NAC 2002)

Figure 1-2: Front cover of *GEC Diesels Australia Ltd* advertisement inserted in *Australian Fisheries* (1976, 35, 8)

Women in the Business

The lack of information collected by fisheries management agencies on the ownership of fishing capital (boats, licenses, quota and businesses) makes it impossible to determine the prevalence of family fishing businesses in the industry and the degree of women's ownership in family fishing businesses. This research provides evidence that the practice of patrilineal inheritance still operates in the fishing industry, particularly in fishing families of European descent. As in farming (Shortall 1999), this practice relies on the construction of fishing as an individualistic occupation and may be rationalised as a mechanism to preserve the family economic unit. Consequences of this practice are that women in fishing often enter the industry through marriage, are less likely than their partners to own fishing capital and less likely to have legal status in their fishing business.

Women's ownership

Ascertaining the degree of women's ownership of fishing capital in this research was difficult due to the many combinations of licenses, quota, businesses and boats with which women were affiliated. While some family fishing businesses are single boat owner operator businesses, many women are involved in family fishing businesses which hold multiple licenses and boats. Several of these businesses are also involved in more than one fishery. An additional confounding factor is that many women are only *partial* legal owners in fishing capital and/or businesses. For example, women may legally be a joint owner of a boat, but not a license, lease or business, or vice versa.

Ownership confers power (Shortall 1999), and the masculine occupational privileging in the industry perpetuates an individualistic and masculine notion of property ownership. This masculine privileging of ownership acts firstly, to confer stakeholder status to the sole or most prominent property owner-the “fisherman”. Secondly, it supports the tradition of patrilineal inheritance, favouring men in the inequitable ownership of fishing capital. Thirdly, it perpetuates the lack of visibility of women in fishing.

> “Women in many of the fisheries, their name isn’t even on the fishing license. So they’re **invisible**. **No body knows they’re there**” (woman, key informant, wild catch sector, in her sixties).

While several women involved in this research recognised that women in fishing, particularly younger women, are increasingly likely to have legal ownership status, women remain less likely than men than to have legal ownership status in their businesses. In this research, the highest prevalence of legal ownership of women was in the oyster fishery. This can be attributed to the sector being established in the 1960s, with the lack of established family fishing businesses meaning that oyster farms tend to be bought rather than inherited. This is in contrast to the rock lobster fishery, established in the 1890s, and Cardea, where fisheries have operated since the 1930s. In both case study sites, these fisheries were primarily established by European families. This has resulted in a strong tradition of patrilineal inheritance and the involvement of extended family members in family fishing businesses.

Certainly, this research confirms it is rare for women to solely own fishing capital. Fisheries management agencies were unable to provide data on women in fishing’s ownership. This failure to collect information on the ownership of fishing capital reflects and perpetuates women’s marginalisation and privileges men in the inequitable ownership of fishing capital.

Women's subjectivities determine the degree of resistance to the masculine privileging of ownership in the fishing industry. Some women who are in second relationships, who are younger and who have tertiary qualifications own other assets in their own name even if they have no legal ownership status in their family fishing business. Some women with no legal ownership status also resist masculine power relations by claiming ownership status, and speak about "my business".

However, many women in fishing with no legal ownership status in their family fishing businesses collude with the dominant masculine industry discourse by locating themselves as supporting and secondary in their businesses, despite making significant contributions.

> "…**they run it and I help them…**the 4 boys are in business together….it **belongs to the boys"** (woman, Cardea, in her fifties).

The pervasiveness of the masculine privileging of ownership is revealed in language which assumes masculine ownership. The following comment reflects resistance by a woman in the oyster fishery whose previous property settlement increased her awareness of her rights to property ownership.

> "The people who really get up my nose the most is the people who talk about **'it's Bill's lease'**, or **'down at Bill's lease'**. I think **'Bloody Hell! Where do I come into that equation?'**…I **do correct them each time** because it does annoy me. So the people locally get the hang of it now" (woman, in partnership, second marriage, in her fifties).

The dominant masculine industry discourse prioritises the family economic unit over the rights of individual women. A further factor in perpetuating the masculine privileging of ownership is the number of women who adopt the legal ownership status most financially beneficial for their family economic unit. Several of these women, with no legal ownership status, explained that the intentional nature of this arrangement was to afford their family fishing business the best taxation advantage, or to absolve the risk of family assets-such as the family home-becoming subject to loss in relation to the significant financial debt, which characterises many family fishing businesses.

This is also the case for some women who are involved in family fishing businesses with collective structures-trusts and companies. These women explained that these structures are created to gain optimum taxation benefits for their businesses. Yet, despite their legal ownership status and reflecting the masculine occupational privileging, in several cases, these women still referred to the business as "*belonging*" to their husband.

Formal regulations may also perpetuate masculine privileging of ownership and the practice of patrilineal inheritance. In some fisheries, it is not legally possible to split fishing assets such as licenses. This was the case in the rock lobster fishery at the time of this research, where there were a minimum number of pots on a license. A woman in an oyster farming partnership explained how previous regulations had stipulated that an abalone license could not be owned in a company name. This had prevented her from entering into a partnership on an abalone license, which her husband still solely owns. Formal regulations and the masculine privileging of ownership in some cases also may exclude women from harvesting fish. For example, one woman who fishes with her husband explained how they owned one license in partnership and another license solely in her husband's name. This woman is only legally able to fish the license held in both names.

Patrilineal inheritance

Patrilineal inheritance practices are prevalent in the industry, particularly in fishing families of European descent. Fisher(men) of European descent tend to be third and fourth generation fisher(men), with the majority having commenced fishing between the ages of 13-16 years and typically involved in family fishing businesses with fathers, uncles, brothers, cousins and sons. These fishing families, who originally migrated to Australia, initially lived in great poverty and worked hard to build up their family fishing businesses. In this context, the prevalence of patrilineal inheritance can be attributed to the need for these individuals to develop economic and social power in order to survive in a new country. Preserving the family economic unit for these families relies on the maintenance of "family" as an extended, non-nuclear collective. This has resulted in a strong tradition of patrilineal inheritance and extended family involvement in culturally specific communities in particular geographic locations such as the Italian community in Cardea.

> "...**Italians were 90% of the blokes that started it off here**....most of the boats here ... their **fathers used to fish** and their **grandfathers used to fish** and now the **sons are running the boats** ...mostly gets handed down..." (male representative, Cardea Fishermen's Cooperative).

Patrilineal inheritance is also evident in some Anglo Saxon fishing families, and aside from ethnicity, there are two further factors which support this practice. Firstly, increasing license fees and operating costs means it is extremely difficult to enter the fishing industry without this financial support from family. Secondly, it is difficult to split fishing licenses and assets. Boats in the wild catch sector are often worth between half a million and a million dollars and are not able to be split. Licensing regulations also mean that certain licenses cannot legally be split. Even where fishing entitlements can legally be split, it may not be economically viable to continue to operate these smaller entitlements. For many women in fishing, the desire to assist their children become independent adults means they support their children (sons) to become established in the industry. This results in families tending to remain in the fishery and passing fishing businesses onto their children (sons) and is a contributing factor to the "closed" nature of the rock lobster fishery and the fisheries in Cardea.

Patrilineal inheritance also reflects and perpetuates the limited options for women to enter into the fishing industry. In this research, the majority of women in the wild catch sector gained entry to the industry through marriage and this was also the case for approximately half of the women in the oyster fishery. The pre-established nature of family businesses and the involvement of extended family members in businesses prior to marriage operate to exclude women from ownership opportunities.

> "… there's my husband, husband's brother…and my father-in-law…..**our names [the wives] aren't on it cause it was established before we got married"** (woman, Cardea, in her thirties).

Discussing succession planning with women reveals the strength of patrilineal inheritance practices and reflects the value attributed to ensuring the survival of the family economic unit, with the family fishing business seen as an investment made for *all children* but as an occupational opportunity for *sons*. Several women in fishing discussed their partners' disappointment when it was not possible to pass their business onto their sons-*"their dream"*-because they either did not have sons or their sons had pursued other careers.

Many women in fishing collude with masculine discursive practices that perpetuate patrilineal inheritance and women's inequitable ownership. One woman in the rock lobster fishery who had two daughters, and two sons who fished in the family business, identified ethnicity as an important determining factor in the masculine privileging of business transmission.

> "It makes it **difficult** when you've got an **ethnic background**. When **you feel a commitment towards your boys**....Unfortunately there's that attitude, 'What would they do?'" (woman, rock lobster fishery, with sons and daughters, in her sixties).

Several women involved in European extended family fishing businesses had not discussed succession planning with their partners and were unaware of how their assets would be passed on. These women *assumed* that the business would eventually be transferred into their *son's name*. Daughters were typically not mentioned in these discussions and several of these women, when asked about their *children* entering the industry mentioned *only their sons*.

Other women in fishing recognised the inequality in patrilineal inheritance practices, particularly where they had daughters only. However, the pervasiveness of patrilineal inheritance is reflected in the justification by many women in fishing that even though men are privileged, they have *worked* for it. This implies that, as daughters do not *work* for these family businesses, they are not entitled to ownership. Several women also expressed the view that this privileging of men is equitable, as women are able to gain qualifications and marry into other established fishing businesses. However, this view fails to recognise that women in fishing do not have an equal opportunity to engage in work for their family business as fishing is constructed as a masculine occupation.

> "Maybe girls … could be a bit resentful because they say the boys get the opportunity to make money more readily and they haven't been given the opportunity....**I think it's fine. I think the boys put in a very, very big workload... so they're really only getting paid for their input**....because I **don't really think it's an industry where girls should be on the deck** somehow.... it's really **a male dominated area**.... I just **don't think it's the place for them"** (woman, rock lobster fishery, ethnic background, two sons who fished and two daughters, in her fifties).

In some instances, and particularly in families of European descent, prioritising the family economic unit results in the delayed transmission of family fishing businesses. Several older women delayed transmission of their businesses due to the possibility that their sons may divorce. This retains power within the original family legal partnership or business and reduces the claim to ownership by women who marry into these families.

One woman with an adult son in the rock lobster fishery explained how she and her husband had retained ownership of their license for this reason. The strength of the collusion of this woman with the dominant masculine industry discourse is evident in her assertion that women wanting to *"go places"* (perhaps referring to women who are assertive, empowered and may threaten the status quo by breaching dominant discursive practices) pose a risk to the family fishing business, and indeed, are responsible for divorce.

> "... **how can you see if things are going to be stable**? People divorce after twenty years. So **you can see what could happen**....This generation there's a lot of it [divorces]. I think **a lot of the women want to go places. That has a lot to do with it"** (woman, rock lobster fishery, European descent, in her sixties).

In some cases, the transmission of family fishing businesses also creates opportunities for women in fishing to become empowered. A woman in Cardea had recently acquired legal ownership status in a family fishing business with her partner. Despite the fact that this woman was going to do the family business book work *"for him"* and with no remuneration, her use of language exposes her altered subjectivity and empowerment in no longer *"**just** being a fisherman's wife"*.

> "We just bought the boat off his parents and his uncle. So my involvement has gone from ...just being a fisherman's wife .... That was his job and his wage, and now I'll be doing the books for him..."(woman, Cardea, one child, in her twenties).

Women's departure

Women's subjectivities evolve over time and some women in fishing, who had unquestioningly entered into personal relationships when they were younger, later wanted to exit these relationships. After having invested 30 years in an oyster partnership, one of these women describes feeling disempowered as the business partnership restricts her capacity to independently invest funds and follow her own goals. While she recognises her option to end the partnership, she also recognises that to do so would have consequences for her personal relationship.

> "I've also looked into financial ways that I can pull the plug if I wanted to...I don't continue down that track because of the **implications that it would have for the partnership and relationship** ....Everything we've done to this stage has been joint. I'm reaching an age where I want to own

an asset in my own right. It's been an evasive thing for me. **I'm a bit like the Polish/Jewish princess who wants to have a diamond to flog if she needs to**" (woman, oyster fishery, tertiary educated, in her fifties).

Another woman in the oyster fishery had migrated to Australia only several months before marrying her husband and had become a business partner for taxation purposes shortly afterwards. She described how her lack of knowledge about the legal and financial implications of becoming a partner had constrained her options when considering separating from her husband and had forced her to remain in the marriage.

"Don said 'It's only for tax purposes so....**I became a partner but I didn't know what I was doing**....I just did it to help Don.... when I was married ten years and had the children...I decide to go cause I couldn't cope any more. So I went to Centrelink...and I said 'What to do?' And she said **'No leave'. So in the end I decide to just keep going**....It's not very nice. And if you haven't got any money, where you go and how do you do it without money?" (woman, oyster fishery, in partnership, three children, in her forties).

Several women acknowledged the increasingly equitable property settlements for women in fishing who had separated from their partners. This greater equality was a result of these women resisting by asserting their claim to ownership.

"In the early fishery, before it really had a value, women had no ownership because the license was in ... the male's name ... If there was a divorce... the woman got the house and the mortgage and the kids.... Guys ... took over the license and the boat ... moved on and became very rich.... Once it got a value and the women became street wise like the males, they just said, 'No, I'm not having half the house...and with my half of the boat and license...[if] I don't believe ... you are delivering your catch ... I will take your half of the license and the boat and lease it to somebody else'. **Women started getting the rights that they deserved**.... since the '90s .... **a few women just started to stand up within the law courts** ..." (woman, rock lobster fishery, in partnership, in her fifties).

However, reflecting the occupational privileging of (fish)men, there were also women in fishing who colluded with the dominant masculine industry discourse, which positioned women who sought an equitable division of assets as being unjust.

"There's a guy ...as legend has it he's probably one of the best cray fisherman ever, and he **doesn't even have his own license**...**his wife's left**

him…he gives her half….that is so sad. The love of his life… and he doesn't even make much money out of it" (woman, rock lobster fishery, in her forties).

Women Who Harvest Fish and Real Fisher(men)

Women who harvest fish

Despite the masculine privileging of the occupation of fishing, many women in this research had harvested fish. These women may be separated into two main categories. The first category accounts for the majority of these women and consists of women who harvest fish with their partners when additional labour is required, often when their family business is experiencing financial hardship. These women are not recognised as having the knowledge or skills required for harvesting fish and instead, like their counterparts in agriculture (Alston 2000), are constructed as a secondary, supporting helper. This construction of women who harvest fish is congruent with the dominant masculine industry discourse, where "fishermen's wives" *do whatever is needed* to support their family businesses and is accepted by the majority of women. As a consequence, as men in fishing are not directly threatened, there is no need for them to derogatorily re/construct these women as abnormal or unnatural. It is also important to acknowledge that many women enjoy harvesting fish and that they may also be able to exercise further power in their family economic units as a result of this discursive compliance.

The second category is comprised of women who assert their right to independently engage in harvesting fish as an occupation and to be recognised for harvesting fish. Ascertaining the number of women who harvest fish as an occupation in the Australian commercial fishing industry is not possible as these statistics are not collected. Women who harvest fish as an occupation breach the dominant masculine industry discourse which define this occupation as masculine and are subsequently derogatory re/constructed as abnormal, unnatural, unfeminine and undesirable. This re/construction typically incorporates aspects of "being a man" embodied in the masculine subject position of "fisherman"-"bloke", tough, dirty, weathered, rough, physical, strong and sexual. The derogatory re/construction of women who assert their right to harvest fish, perpetuates masculine advantage by establishing a further barrier for women to enter the fishing industry.

Variations in location, natural environments and fishing practices have differing implications for women's capacity to engage in harvesting. For example, several women commented on the increased opportunity for and acceptability of women to harvest rock lobsters at the Abrolhos Islands where reefs are shallow. The capacity of women to resist exclusionary practices is also influenced by their up bringing, including exposure to discourses of equality, being subject to physically active or non traditional female role models and previous experience in fishing or farming, often alongside their fathers. Resistance was also facilitated for some women by exposure to alternate gender discourses through tertiary study and one woman explained that this was why she would be happy for her daughters to fish. Again, the strength of the dominant masculine industry discourse is revealed when she indicates that her husband would have a problem with this. As is common in research with women in farming (eg, Grace and Lennie 1998; Teather 1994a), the term "feminism" was only used by one woman in this research. However, feminist sentiments are strong in the subjectivities of some women in fishing.

> "There are a few women skippers and deckhands. ... women could actually work their way through it, then you expect it. But this is just, **we are on the outskirts, this war is going to stay there**, if you work on a boat you must be **bullish**. There must be something **abnormal** about you" (woman, rock lobster fishery, in her forties).

The likelihood that women harvest fish, and the degree of derogatory re/constructing of these women, is also influenced by ethnicity. In Cardea, the re/construction and ostracism of women who harvested fish was particularly harsh, with this being the only case study in this research where no women had participated in harvesting fish. This may be attributed to the influence of European culture in this specific area in two central ways. Firstly, the necessity for women in fishing to contribute their labour in times of financial hardship is reduced due to the involvement of extended family members in fishing businesses. Secondly, ethnicity in this research correlates with more rigid "mothering" and "wifing" discourses. This results in women in Cardea more stringently adopting/ being relegated the subject position of "fisherman's wife" and the greater uptake of associated acceptable practices.

The pervasiveness of women's exclusion from harvesting fish was revealed through discussions with women on how their partners felt about their daughters fishing. Collusion by both individual fisherman fathers and fishermen collectively confines and encourages their daughters to adopt acceptable practices dictated by broader “wifing” and “mothering” discourses.

> “… because of the **Continental background** …. .it is really difficult for the guys to allow their daughters to be individuals….I think it would be great [if my daughter wanted to fish]….but the **rest of the guys** down at the wharf **might have a problem** with it….**Serge [husband] would have the biggest problem** because **girls just don’t do that…**Because **girls stay home and cook meals and have babies”** (woman, Cardea, with sons who fished, in her fifties).

While this woman from Cardea indicates her potential support for her daughter to fish, the majority of women in Cardea collude with their partners in excluding their daughters from fishing. These women perpetuate the dominant discursive re/construction of women who harvest fish as abnormal and unnatural *based solely* on the gendered construction of “woman”-fishing is “*not a good thing for women”* and “*not a woman’s job*”.

> “I don’t think she [my daughter] would [want to fish]….she’s **not the type**, she wouldn’t.…[my husband would] **probably talk her out of it** cause …**women going fishing, it’s not a good thing”** (woman, Cardea, with sons who fished, in her sixties).

This practice also demonstrates how these discursive re/constructions and the masculine occupational privileging of fishing are perpetuated and transferred between *generations*. The subtlety of this discursive maintenance is also evident in the following extract, where a woman from Cardea espouses this view while simultaneously asserting she is not sexist.

> “There would be **no way** [my husband would allow our daughter to fish]. No….it’s **just not a woman’s job**, and **I don’t mean to sound sexist** saying that” (woman, Cardea, with sons who fished, in fifties).

Real (fisher)men and men’s work

Fishing is a dangerous occupation and the high accident and fatality rate (ILO 1999) is attributable to the harsh, intense and unpredictable nature of the environment. Fishing is constructed as “men’s work”. “Fisherman”, in turn, is a subject position in which occupants require and possess qualities associated with traditional notions of masculinity-heroism, bravery, risk taking, strength and toughness; and engage in "masculine" practices, including hunting, larrikinism and collective drinking.

Hunting in the face of danger

"Fishermen" are constructed as *"hunters"* and one description of "fishermen" as *"the last of the wild hunters"* also alludes to the embodied nature of this hunting instinct. Hunting, in turn, is constructed as a masculine activity and as one enjoyable aspect of *"being a bloke"*.

> "I guess most **blokes like that hunter gatherer stuff** don't they. A lot of blokes do" (woman in fishing, key informant).

Fishing requires bravery to face environmental conditions and "fishermen" are often constructed as having an intimate and mystical relationship with the weather and ocean. The relationship between fishermen and nature has also historically invoked an association between fishermen and God, particularly for ethnic fishermen. This is evident in descriptions of Blessing of the Fleet ceremonies, a traditional European celebration where boats are blessed at the commencement of the fishing season. This celebration is held in Cardea and in other areas in Australia where there are a high percentage of European fishermen.

> "We ask you to bless the **fishermen** of this fleet. Keep them safe from the **dangers of wind and waves**....bring them safely through the **trials and dangers of the sea....O God of majesty, for those in peril** on the sea.... **preserve their lives in danger's hour**..." (prayer extract from the 2003 Blessing of the Fleet brochure, Cardea).

This relationship between fishermen, nature and God elevates the heroic and masculine qualities embodied in and required by fishermen to face danger. This reflects and perpetuates masculine occupational privileging and excludes women, who are not constructed as possessing the necessary masculine and heroic qualities required for fishing.

> "It just seems to be one of those **last bastions of male domination**... perhaps it **stems from the danger** of going out to sea and it was considered as **a man only job**..." (woman in fishing, key informant, wild catch).

> "...**dangerous**....I mean, it's still such a **man's industry**. I don't know who in their right mind would want to go out there!" (woman, rock lobster fishery, in her thirties).

Hard and dirty work

Harvesting fish is physically demanding and dirty work, with these aspects of work traditionally associated with "men's work". In this research, many women in fishing expressed the view that women lack the required physical capacity to harvest fish.

> "I mean we've got no female deckhands around here ….we've had a couple of girls ask for jobs, Sergio says, '**you can't expect a woman to lift'.** There's a **lot of lifting** and its **hard and physical labour**…" (woman, Cardea, in her sixties).

Some women recognised the capacity for women in fishing to increase their physical contributions if harvesting tasks were conducted and designed "*differently*", for example, lifting with two people or employing mechanisation. One woman described how many women work physically with their partners in a supporting role in her fishery, but are restricted in becoming sole owner operators due to the physical demands of the job, particularly as bodies age.

> "…it's physically fairly **difficult for a woman** to do that **on her own**…. It would be very challenging…**Not saying you couldn't do it**….You'd probably do it slower and you'd **certainly do it differently to what blokes do** ….Now if my husband couldn't go…**I could probably do it**, but I would find that quite **difficult**, **especially now I'm older"** (women in fishing, wild catch, key informant, in her sixties).

Another woman who was involved in industry decision making felt that the aquaculture sector provides increased opportunities for women's equitable involvement in harvesting fish due to firstly, lesser perceived physical demands than in the wild catch sector and secondly, the more recent nature of the aquaculture sector, where the traditional gendered division of labour may not be so entrenched.

> "…a lot of it's land based, and the **duties are not so physically demanding compared to wild catch fishing**…so I actually see **women playing more of a role** in that, and able to be involved in the physical running of the farm… we are talking a **true partnership**. And **sharing of duties that you don't see in the wild fishing"** (woman in fishing, wild catch, key informant, tertiary educated in her thirties).

However, for many women in fishing, potential opportunities for them to engage in physical labour are constrained by discourses of gendered embodied difference. That is, women's "natural" tendencies towards, preference for and adeptness at "cleanliness", hygiene, attention to detail, observation and nurturing.

> ".... a lot of **women don't want to be lifting** bags all day **and trudging around in the mud**....there's a **preference for...cleaner** type work....hygiene, quality control, attention to detail and observation...it tends to be a **natural tendency** that **those qualities are a bit stronger in women"** (woman in fishing, aquaculture, key informant, tertiary educated, in her forties).

In some cases, women's alleged "natural" embodied qualities, including patience and a greater capacity to tolerate boring work, function to relegate women to routine, repetitive tasks. One of these tasks is grading oysters, which occurs post harvest and involves sorting the oysters according to size. This task is undertaken by many women in the oyster fishery after their partners have harvested the oysters.

> **"Women are able to cope with mundane boring jobs. Men can't**....Women can stand there and grade oysters like that and think about what they have to get done at home that day or plan their weeks grocery menu" (woman, *oyster fishery,* in her thirties).

While fishing can be a dirty occupation, women who harvested fish were re/constructed as having lesser personal hygiene standards than other women. Being "unclean" is considered undesirable or inappropriate for women and operates as an exclusionary mechanism to prevent women gaining access to the masculine preserve of harvesting fish.

> "...[my daughter fishing was] **out of the question. Dad** always wanted her to work in the chemist shop**. Nice clean job"** (woman, Cardea, with sons who fished, in her fifties).

Drinking

While several women in fishing suggested there was a serious level of drug use in the industry, specifically marijuana and heroin, many more women spoke about a culture of drinking alcohol, a practice which reinforces the collective masculine "fisherman" identity. Indeed, an indication of the cultural pervasiveness of this behaviour is demonstrated by its inclusion in the *Eastern Tuna and Billfish Fishery Industry Code of Practice* (Jusseit and Robinson 2003).

> "...fishers should act responsibly to maintain the professional image of the ETBF. In particular **fishers should....avoid drunk and disorderly behaviour in public places"** (Jusseit and Robinson 2003, 26).

Reference to these "public places" indicates that collective drinking by fishermen occurs in particular spaces. Indeed, this relates to a historically pervasive gendered spatial occupation of specific places, with fishermen drinking on boats, wharfs and pubs. These are public visible spaces, where historically fishermen have gathered to conduct fishing business while women remain at home in private spaces.

> "No, **wives don't** tend to **come to the hotel**....I think the Italian wife is a home loving person and they seem to **stay at home** and **look after the family** and the **meals for the old men"** (interview extract with the Cardea publican from Cardea documentary, 1970s).

> "Women rarely went into the public bar even though they were allowed, it **really was the guys 'domain'** and was many times **too rough"** (local history book Cardea).

This "roughness" also alludes to how the practice of collective drinking provides fishermen with an opportunity to reinforce their masculinity by engaging in masculine practices such as fighting and telling "tales".

> "The fishermen drank at the pub and **told tales** about the catch.... The [Cardea **pub] was the fishermen's second home**. There was many a **fight and scuffle**, but that was **just part of the life"** (local history book Cardea).

> "...you hear all sorts of tales. A mate reckons he worked for a bloke over the islands used to pour **bloody beer on his cornflakes in the morning**. Someone else ate goldfish, shit like that. Another friend's father had a wooden leg and died out in the channel here. Then there's the lie....The **most crays ever been caught** in the world is from **the front bar in the Gero. pub!"** (interview extract with rock lobster fisherman, de Monchaux 1999, 161)

In some instances, the drinking culture also incorporates an overt masculine sexuality with fishermen displaying parts of their bodies which are usually covered. This sexualised behaviour is associated with specific physical locations where drinking occurred, quite commonly in pubs.

> "I had a **habit** of **taking me clothes off**....Have you ever seen a bulldog? ....You'd get down on your hands and knees, **stick your balls out**

> **between your arse then run across the bar"** (interview extract with fisherman, de Monchaux 1999, 20).

An interview with a barmaid, who works at a tavern frequented by fishermen, colludes with this reinforcement of dominant masculinity by finding it humorous. Furthermore, while she is accepted in this masculine space, her secondary status is evident in her work being contained in a defined physical space-behind the bar-and in her role as serving the fishermen.

> **"The funniest thing?** Well I've seen **heaps of penises** [laughs]...by all these guys **getting dacked in the bar in front of us**....Bully must be the hero for dacking the most people... I saw a **pair of boxers being burnt** in the corner over there the other night, that was pretty funny [laughs]..." (interview extract from Shorewater magazine 2003).

The pervasiveness of collective drinking is indicated when a rock lobster woman describes having a partner who does not go to the pub as a "*luxury*". This also reflects the diversity in masculinities engendered in fishermen.

> "I'm pretty **lucky**. My husband...likes to come home each day to me. He doesn't want to go to the pub...But **not so many other women have that kind of luxury"** (woman, rock lobster fishery, three children, in her twenties).

While there is evidence that certain sites are imbued with masculine power as locations for drinking, the following poem written by a rock lobster woman suggests that what excludes women is not the physical *site* where drinking occurs *per se*, but the collective act of drinking exclusively with other fishermen (*us*), *characterised* by the *absence* of women partners (*them*).

COCK STROKE

Hey mate
Aren't we great
Liftin' crates
6 months? No! 8!

Missus' at home
She made a fuss
Whinge and moan
That's all she does

Celebrate with her? No way
She's hung around 8 months already
What's another night or day
I'll get trashed with Mike and Eddy

Didn't tell her 'bout the piss-up
Or share the excitement 'bout the wind-up
Us blokes, we don't tell "them" our plans
Makes us feel not quite a man

(Poem by Ruby from Cervantes, in Shorewater magazine 2003, 13)

Women's Exclusion from the Practice of Harvesting Fish

Women are actively excluded from the practice of harvesting fish. The culture of particular fisheries and locations determines the degree of masculinisation of specific discursive sites of power. Given the geographical dispersion of farms in the Tasmanian oyster fishery, boats, wharves or jetties are more likely not to be visible public spaces, and thus, are not discursive sites of masculine power to the same extent as in the wild catch sector. All of the women involved in this research in the oyster fishery had contributed manually on their farms. However, the majority of these women remain invisible as they undertake this manual work in geographically dispersed locations in the capacity of secondary "helpers". Indeed, this discursive positioning of women in fishing is so entrenched that, while all of the women in the oyster fishery have worked on their farms, no woman who was interviewed believed that any other woman in the oyster fishery is also engaged in manual labour.

In contrast, in the wild catch sector, boats and wharves are primary and public sites of dominant masculine power. Boats are often feminised and fishermen are described as having a personal and intimate relationship with their boats.

> "I changed the name to **'Sea Wyf'** (you're **married to your boat** once you've got one)" (interview extract with fishermen, de Monchaux 1999, 56).

> "… they [fishermen] have a **love affair with their boat**. It's **really their mistress"** (woman in fishing, key informant, wild catch, in her fifties).

The size of boats also conveyed masculine power which, in the following case, is accompanied by a corresponding masculine sexuality.

> "My dad built the **biggest boat** and what did we call it? **His phallic symbol!**....[it] costs him heaps in fuel to get to places....and it's a waste of time cause it's a slow beast....Heaps of ego!" (woman, rock lobster fishery, three children, in her twenties).

There are significant discursive regulatory practices, which operate to exclude women from boats and wharves as sites of masculine power and preserve the practice of fishing as masculine. These barriers include beliefs that the personal hygiene needs of women are not able to be met on boats due to the lack of adequate facilities; the superstition that it is bad luck for women to be on boats; women's assumed embodied propensity for seasickness; the sexualisation of sites of power where women on boats are sexually harassed and constructed as a sexual distraction to fishermen; and women's lack of opportunity to engage in physical work and develop the strength required to harvest fish.

A lack of facilities

Despite the fact that newer boats had hygiene facilities such as showers and toilets, many women in fishing explained that the lack of these facilities is one reason why women do not harvest fish. However, the following extract suggests that it is not the lack of facilities per se which exclude women, but a masculine appropriation of the boat and the behaviour of the men who harvest fish.

> "...you think about a woman...having to go down stairs and having a shower **when all those guys are down there** and...**it wasn't sort of on**.... the **fishermen are a rough lot** most of them..." (woman, Cardea, with sons who fished, in her fifties).

During the course of this research, I met another woman doing a fisheries related doctorate and we became friends. One evening she described her fieldwork experience of being the only woman on a fishing boat for trips of several days at a time. There was no toilet, and the crew urinated and defecated off the side of the boat. There was nowhere to do this which was not visible from at least one other area of the boat and, in jest, the fishermen regularly took video footage of other crew members defecating.

Where subjugated knowledge of (fisher)men's experience is revealed, it is not dissimilar to the views expressed by many women in fishing, indicating these practices also function to exclude men who do not embody the required masculine fisherman identity.

> "I suddenly became paranoid about needing to go to the loo. I **can not hang over the rail like the rest of them**. I would **rather leap into shark-infested waters** and **swim back to the bush**" (interview extract with fisherman, de Monchaux 1999, 65).

The discursive absence of (fisher)men's experience in relation to the lack of hygiene facilities provides evidence that the gendered embodied assumption that women require their bodies to be cleaner than men's, is a gendered exclusionary mechanism. (Fisher)men may be unclean and defecate in public, but this is seen as "unnatural", undesirable and inappropriate for women. These gendered assumptions and exclusionary practices operate to preserve boats as sites of masculine power and preserve the practice of harvesting for men.

"It's bad luck to have them on the boat"

All of the women from Cardea and several other women from diverse cultural backgrounds in the wild catch sector discussed the superstition that it was bad luck to have women on boats. The majority of these women felt that their partners personally did not hold these beliefs. However, language such as "not *allowed* on the boat" reflects its potency for excluding women, thereby preserving boats and the practice of harvesting fish as masculine domains and perpetuating the masculine privileging of harvesting fish as a man's occupation.

> "women **aren't even allowed on the boat**… it's considered to be **a man's business"** (woman in fishing, key informant, wild catch).

> "…people seem to think that we have…this religion that … it's bad luck to have girls on the boat, but that really isn't the case. The **idea is that it's not a place for girls**…. it was a **man's world out there**, and **they didn't like to take women for that reason**. **Not to be exposed** to that sort of behaviour…" (woman, Cardea, in her fifties).

This superstition functions to prevent women from gaining the skills and knowledge of the practice of harvesting fish. This becomes apparent in the story of a woman from Cardea, whose husband was happy to take female research scientists on the boat but not women who wanted to harvest fish. In this case, the subject position of "scientist" does not have the same potential to appropriate (fisher)men's skills, knowledge and power as a (fisher)man's partner could, should she begin to harvest fish.

> Cardea woman: "Even if I wanted to [go fishing] I am a woman, its bad luck to take a woman out. Excuse me, but **what did I say before about all these female research scientists**?"
> Jane: "So what's the difference?"
> Cardea woman: **"I'm married to him"** (laughs) (woman, Cardea, in her fifties).

These women did not harvest fish and occasionally went on the boat to provide company to their partners. One younger woman from Cardea resisted women's exclusion by threatening to eject a male from the boat, albeit a younger male.

> "His little cousin said to me **'What are you doing on the boat? You're bad luck'** I said **'I'll throw you over the side. That would be bad luck**!'" (woman, Cardea, in her early twenties).

Seasickness

Seasickness is not a gender specific condition. However, seasickness was constructed in a gendered embodied dichotomy of proven male strength and assumed female weakness, with the majority of women in this research who do not harvest fish explaining that seasickness is the reason why they do not go on the boat and/ or why they *"are not taken"*. That is, why they are excluded from harvesting fish.

> "They **won't take me**... because they know how bad I am" (laughs) (woman, Cardea, in her forties).

Furthermore, the potency of seasickness as an exclusionary mechanism is demonstrated with many women in fishing either not engaging in, or not being allowed to engage in, the practice of harvesting fish based *solely* on the *assumption* of their propensity towards seasickness. Despite many women in fishing expressing an interest in going on the boat, they collude by accepting they will probably be sick.

> Jane: "Do you want to go fishing?"
> Cardea woman: "I'd **probably get sea sick**....I've said to him once I wouldn't mind going out just to see what they do. So what they did was took a video camera of what they did. So that was good" (laughs) (woman, Cardea, in her forties).

Men also actively employ discursive practices to reinforce seasickness as embodied in women. In the following extract, the language used reflects the masculine subject position of "fishermen", with the horrific test forced onto *women* indicating the power of "fishermen" to dominate and preserve boats as their own domain.

> "Bronte:....On the back of the door in the kitchen was this graph … on the amount of girls they would take out and how many would be sick.
> Paul: …. All the guys on the boat, we all take girls out, tourists…and they had to **be sick to get points** and at the end of the year … whoever had the most points, had to **buy kegs**. It was an **honour system with the guys**.…we used to use cows legs in those days [for bait] … and sometimes these girls…we couldn't get them to be sick. You know, we tried everything and the **bastards wouldn't be sick** …. 'You'd like something to eat girls?' ….And they'd look up and there's this **rotten cow's leg** in between two bits of bread and that would usually do it, you know – **they'd be out the door and spewing"** (France, Mitchell and Wright 1998, 149).

(Fisher)men's experience of seasickness is also peripheral in the discourse. Where this subjugated knowledge is revealed, it appears that (fisher)men also get seasick but have the *opportunity* to overcome it. The only woman in this research to mention her partner experiencing seasickness when he commenced fishing described how he overcame it with determination and continually going on the boat. In some cases, this "opportunity" to overcome seasickness was forced upon men in fishing as young children.

> "Dad used to take us out **whether we liked it or not** and me and Raymond, my twin .... We'd get seasick from the time we left the heads to the time we got home, then he'd take us out the next day, and the next day, and **eventually we got used to it"** (interview extract with rock lobster fisherman, de Monchaux 1999, 225).

Sexualised sites of power

In the wild catch sector, the two main sites of masculine power, boats and wharves, are sexualised. Women are excluded from these sites by two pervasive discursive practices-the re/construction of women who harvest fish as innately sexual, and by sexual harassment and assault. Many women experience or are aware of incidences of sexual harassment and assault, and are aware of the pervasiveness of this culture on wild catch boats.

"…some skippers just think well **you're only there for one reason** and **that's to go to bed with them** and that's it" (woman who harvests fish, wild catch, in her forties).

"…. this deckie wanted to take a few girls out with him… and took an esky of beer….they were **rubbing their hands together** thinking of what was going to happen out there….I think girls don't know that they have a **meter box in the cabin where they put up a whatsie [tally to indicate sexual encounters** with women]…. I think they thought they were gentlemen, but **it's a sex thing"** (woman, rock lobster fishery, in her forties).

This behaviour also occurs on wharves.

"I used to sit up at the bilge with my daughter…and the guys next door were going on about their wives…**guy talk**… but they don't worry about saying things to girls on the jetty. **If you go down onto the jetty, you are just an open target"** (woman, rock lobster fishery, in her forties).

This sexualised workplace culture operates to exclude women from the practice of harvesting fish and from boats and wharves in three main ways. Firstly, it reduces the likelihood that parents allow or encourage their daughters to harvest fish.

"I **wouldn't like my daughter to do it** [fish]….**All the guys sleep with them"** (woman, rock lobster fishery, European, with a son who fished, in her sixties).

Secondly, women are less likely to be employed on boats. Two male representatives from industry organisations, who are also skippers, described how they do not take women on boats as this "distraction" would interfere with the crew harvesting fish. One woman, who is widowed and solely owned her rock lobster business, explained she would not employ a woman skipper for this reason.

"… they'd have to **handle the guys out there jumping onto you** and don't think they don't! They get **targeted as soon as they got out there**. It would have to be an **exceptionally strong woman**, mentally, and they'd have to be able to cope with that" (woman, rock lobster fishery, sole owner, widowed, in her forties).

Thirdly, in the wild catch sector, and particularly in Cardea, many women in fishing insist that their partners do not employ women to fish with them.

> "… one thing I was adamant about, some girls used to go and ask him for jobs. But **I said to him**…'I'm not trying to tell you what to do, but I don't want women on the boat**. I'm the only one that goes fishing on that boat**. There's not going to be other women on there'….it would have caused **problems with the crew** ….you've got to be a **certain type** of person to do that…" (woman, Cardea, in her fifties).

Excluding women who potentially may be involved in a breach of monogamy from harvesting fish with their partners can be explained by the desire of women to preserve their family economic units, and their power within these units. Many fishermen also comply with this (at least overtly), which is perhaps attributable to (fisher)men's reliance on women to maintain the family and family fishing business. Women's own capacity to provide or withhold sex also enables women to exercise power in this situation. Ironically, this collusive view held by the majority of women in fishing that women should not harvest fish is at the expense of women who do, and who want to, harvest fish. Further evidence for this argument is demonstrated with the sexual construction of women who fish *not* applying to women who fish with their partners. Indeed, one woman felt that fishing with her husband protected her reputation.

The options for women to resist are constrained by the potency of these constructions. Often women who harvest fish need to employ strategies to enter into and remain in this masculine preserve. One woman explained that "*sleeping with the skipper*" is a strategy employed by some women to gain access; and another woman who fished described how she intentionally acts "*like a man*" to avoid sexual harassment.

> "I like try and **act a bit more mannish** … like you **swear and carry on**. You **don't want to be perceived like that"** [in a sexual manner] (woman who harvests rock lobster, in her thirties).

Thus, the discursive practices which operate to exclude women from harvesting fish and the strategies adopted by women who do enter this preserve, perpetuate and reinforce the re/construction of women who harvest fish as abnormal, unnatural, "like a man" and sexual.

Physicality of gear/ boat maintenance and harvesting fish

Women are excluded from physical work on boats at wharves. This can be explained by (fisher)men not wanting women to appropriate their claim to harvesting fish as "men's work", particularly in a visible and public site. Policing of this occurs collectively by fisher(men) at the wharf.

> "...he was pretty tight for cash....We were doing as much as we could and then **he started getting feedback that the guys were going 'no, no, no'**....There are times when I know that he needs a hand and **I'd love to be down there** helping him, **but it's just, you don't go"** (woman, rock lobster fishery, in second marriage, in her fifties).

Women in fishing also ostracise women who breach these discursive boundaries.

> "Paul used to ask me if I could go and give him a hand. It was nothing for me to go on the boat ...giving him a hand to pull the engine to bits .... I'd been doing that on the farm...At first it didn't mean anything for me to just do that. And then slowly and surely **the women kept saying 'You shouldn't be doing that**..."(woman, rock lobster fishery, in her fifties)

This may be explained by women wanting to reinforce their own identity as "fisherman's wife" and not wanting to risk ostracism themselves.

Lifting pots

Certain tasks in specific fisheries are extremely physically demanding such as lifting pots in the rock lobster fishery. These pots weigh approximately 70kgs each and deckhands lift and stack many pots on the boat each day. Lifting pots was referred to repeatedly during research in the rock lobster fishery as one task which "women" were not physically capable of. Indeed, there was only woman I learned of who had done this.

Reflecting the masculine privileging of this task is the discursive absence of men's experiences of the physical challenge of lifting and stacking pots. A subjugated masculine knowledge is revealed in the following extract, which indicates the difficulty in training *a man's body* for the task (*"insanity"*) of lifting pots and the cost of being seen as *"soft"* and not a *"real man*" for finding it difficult or acknowledging the physically challenging nature of the task.

> "When I first lifted a cray pot I thought – fuck this. I could barely lift it off the ground. What is this insanity? After three years of university pen pushing I knew, looking at the stack of a hundred-and-twenty-three craypots in front of me; here were some objects that were going to cause me anguish and pain. I started out handling only a few every day. The

> other deckhands watched in amusement as I groveled the pots about the deck looking like I was about to cry…because I was about to cry. I didn't have the strength! …I looked at the deckys pleadingly. Help! But no help was forthcoming. I had to deal with pots all by myself. There was no escape….My workmates amusement soon turned to frustration and resentment … my skipper would look at me, scowling….I would avoid his glare, feeling ashamed of my total inability to handle the pots. Eventually he'd just scream at me, "Fucken' hurry up!" ….I just couldn't get on top of pots. They **completely disempowered me as an individual amongst my fellow deckhanding comrades**. There was **no way I could handle a full man's load**, and hence there was **no way I could be considered a real man.** I didn't really cut the mustard. **I was soft"** (interview extract with fisherman, de Monchaux 1999, 48).

This subjugated masculine knowledge reveals lifting pots as a test of "manhood". In this context, "women" can never be "real men" no matter how many pots they lift. This test of manhood, combined with discourse which constructs women as inherently lacking the physical capacity for harvesting fish, functions to exclude women from having an equal opportunity to attempt to train their bodies over the same time period.

> "Because fishing is such a **physically oriented industry**, **men will always dominate**…" (woman, rock lobster fishery, in her forties).

Furthermore, it also highlights the diversity amongst men. There are men who would not be considered as "real men", i.e., men who would not have the required physical capacity and presumably, some who would not want to try

It was recognised by some women in fishing that altering the method of lifting and stacking pots could result in the increased participation of women. Indeed, women are supportive of making changes to reduce the physical demands on *all* bodies. However, there is resistance from rock lobster fisher(men) over attempts to improve occupational health and safety in relation to lifting pots. This resistance could reflect genuine concerns with impractical regulations impeding harvesting practises. However, it is also interesting that this resistance is occurring over the most potentially physically damaging task involved in rock lobster fishing for *all bodies*-the task which defines "real fishermen" and excludes others, including women and "soft" men.

Changing bodily forms

In the rock lobster fishery and other fisheries, women are further disadvantaged as the physical work of fishing and the exposure to the elements is often accompanied by changing bodily form. Fisher(men) have privileged opportunities for developing strength, a desirable masculine quality, while women who develop muscular, strong bodies are derogatorily re/constructed as "*like a man*"-undesirable, unfeminine, weathered and "unnatural".

> Patty: "Most of the women who fish **are** pretty sort of…"
> Bill: "**Blokes"**
> Patty: "Yeah. They're **not particularly feminine"**
> Bill: "They're not feminine…they're **pretty brutal looking"** (Bill and Patty laugh)
> Patty: "And people become **pretty weathered** when you're out there as well" (married couple, rock lobster fishery, in partnership, with daughters who do not fish, in their forties).

Deconstruction reveals the fallacy of this myth. One woman who harvested fish with her husband explained how she was seen as an enigma, because she did not have the physical appearance, or display masculine behaviours such as swearing, to match this masculine discursive re/construction of abnormal, unnatural and "tough" fishing woman.

> "…because I **don't swear** and I **don't have tats** and I'm **not a big buxom broad** sort of person" (woman in fishing, wild catch, in her fifties).

The level of hostility in the re/construction of women who harvested fish as abnormal and unnatural was also mediated by personal relationships. One woman described a woman rock lobster skipper she knows as being "six foot tall and built bigger than the guys". While this skipper is constructed as embodying masculinity in her size and build, she is also described as "really nice". This reveals the conflict inherent between the derogatory re/construction of women who harvest fish as abnormal and unnatural, and the personal experience of knowing a woman who fishes.

The exclusionary power of the derogatory physical construction of women who harvest fish and the assumption that women lack the required or appropriate physical bodies for fishing is evident.

> "I mean, there are women who are skippers but **have you ever seen them**?....I **wouldn't like my daughter to do it** ….I mean, **they're really**

rough" (woman, rock lobster fishery, European, son who fishes, in her sixties).

Fishermen's Knowledge – in the Blood and in the Know

The construction of fishing as a masculine individual occupation, patrilineal inheritance and regulatory practices which exclude women from the practice of fishing, all contribute to the commonly expressed belief that (fisher)men "know" fishing in a way which women in fishing do not. This privileged construction of (fisher)men's embodied knowledge typically incorporates a unique and innate connection with the sea and the weather. Language used repeatedly to refer to this included fishermen's *"intuition"* and *"natural feel for the sea"*, with fishermen having *"sea water for blood"*, *"sea water in their veins"*, and fishing *"in their blood"*. One woman in fishing colluded with this discursive masculine construction of "fishermen's" connection with harvesting fish and the sea by making no reference to her feeling about the sea, despite the fact that she also lives on board the boat and fishes with her husband.

> "… [fishermen have] got **sea water for blood**….Pete has said, many a time … **'Something's telling me…we've got to go in'**. And sure enough we'll go in and have a breakdown or something …" (woman in fishing, wild catch, in her fifties).

However, it can be suggested that fishermen gain their knowledge (skills) through practical experience, in many cases through the privileged opportunity afforded by patrilineal inheritance.

> "Well a lot of them are **fourth, fifth generation fishers**, so I guess that's **born into them"** (woman in fishing, wild catch, key informant).

> "… young men … said goodbye to their Italian motherland…. Even though they were strangers to this new land, they were **not foreigners to the sea**. They were their **father's sons**, of the lineage where **tradition and salt water flowed through their veins** – where **fisher boys grow into fishermen"** (Mayer 1997, 15).

"Knowledge" may therefore be seen to relate to practical experience. One woman in the rock lobster fishery makes the significant distinction between skill and knowledge.

> "I don't think that anyone could say that they could catch a rock lobster if they didn't go out and my husband can go out and put his pots … smack on

the edge of the coral … by looking over the side and I couldn't do that, but that **kind of skill I guess comes with practice"** (woman, rock lobster fishery, in her forties).

Another woman in fishing demonstrated the pervasiveness of the dominant discourse by being grateful to her husband for sharing his knowledge and commenting on the surprise of his deckhands that he had done this. This is despite the fact that she inherited the license from her father and has retained sole legal ownership while her husband skippers their boat.

"John … **shows me** the best places to go to and best place to put it [the pot]… it **spins the other crew out** as well. That's because **my husband's pretty good to me"** (woman, rock lobster fishery, three children, in her twenties).

Several women in fishing resisted the notion that women could not gain the skills to "know fishing". While some women did claim this knowledge, in most cases this knowledge was gained through secondary sources and not from practical fishing experience.

"… **women do actually know** what is going on ….**I know** all the time where my husband is…. how many crays they caught out of what pot …where they were moving, where they were getting their crays and how the crays change from one season to the next and what the catch was 5 years ago" (woman, rock lobster fishery, in her sixties).

However, many women in fishing colluded with the construction of "fishermen's" knowledge as being contingent on practical harvesting experience. This was also reflected in the majority of women not claiming fishing knowledge, despite being able to comprehensively discuss both many aspects of fishing practices and current industry issues.

"…**a lot of women… don't understand because they haven't been in a boat**… If you understand what happens in a boat you can understand the finer details…." (woman who had worked as a deckhand, rock lobster fishery, three children, in her forties).

The masculine privileging of harvesting fish means women in fishing do not have the same opportunity to practice fishing skills and claim the knowledge of (fisher)men. In turn, this lack of knowledge justifies women's exclusion and perpetuates the occupation of fishing as a masculine domain. The dominant masculine industry discourse is that women's lack of knowing means they do not have the knowledge to participate, or be recognised, in the industry.

> "... [women's] opinions are not really valued...the fishermen think **'Well we don't catch fish**, so **we don't really know** what is going on out there', and it's **just a male dominated industry"** (woman, rock lobster fishery, in her forties).

Conclusion

The Australian commercial fishing industry is characterised by a dominant masculine industry discourse which constructs fishing as an individualistic masculine occupation undertaken by men-"fishermen"-who are involved in the practical aspects of harvesting in public visible space.

> "...at the moment there's **not a huge number of women** who actually go **on boats** and **that's what people think of as the fishing industry"** (woman in fishing, wild catch, key informant).

However, this construction of the industry does not acknowledge the multitude of tasks involved in operating a family fishing business and renders women in fishing invisible by failing to recognise their substantial contributions. It also must be recognised that power, and subject positions, are productive of identity (Foucault 1980), and that some women in fishing experience personal fulfillment, an internal source of empowerment (Deveaux 1994), as a result of doing what they do. However, the daily experiences of many women reveal that the dominant masculine industry discourse is pervasive and has material consequences (Hartsock 2004a).

The industry is imbued with regulatory discursive practices, which both reflect and perpetuate dominant masculine power relations, defining and privileging fishing as "men's work" and excluding women from the practice of harvesting fish. Women are excluded from masculine sites of power, boats and wharves, through various discursive practices. Women's marginalisation is also reflected and perpetuated in patrilineal inheritance practices, which limits women's entry into the fishing industry to marriage and reduce opportunities for women's ownership of fishing capital. Fishermen are also constructed as possessing a peculiarly masculine "blokey" subjectivity which is necessary for harvesting fish-tough, dirty, risky "men's work", and which women lack.

"Fishermen" are also alleged to have a unique Gift, with fishing "in their blood" or "sea water in their veins", and as a result only they can "know" fishing. While there can be no doubt that fishermen gain experiential knowledge through their daily activities (Maurstad 2000; Palsson 2000), this masculine knowledge/ power fantasy claims legitimacy through the (alleged) embodied "natural" affinity of (fisher)men with both the practice of harvesting and particular constructions of nature. Yet, this knowledge may be attributed to the masculine privileging of the opportunity to practice a specialised skill through the tradition of patrilineal inheritance. Women's exclusion from fishing prevents women from accessing the opportunities to "know" as (fisher)men do and perpetuates masculine occupational privileging in the industry. In turn, women's absence from fishing is commonly justified by their lack of knowledge.

While men in fishing are privileged with an occupational status ("fishermen"), women in fishing are positioned as secondary and identified by their marital status ("fishermen's wives"). Acceptable activities for women in fishing are based on constructions of bodily difference and feminine subjectivities about what is "natural" for women to do, such as support, care and nurture, evident in discourses of "wifing" and "mothering". The dominant masculine industry discourse incorporates these more generic discourses and constructs women as lacking the physical capacity required for harvesting fish, and fishing as an undesirable occupation for women.

However, approximately half of the women involved in this research, at some point, have harvested fish. In the wild catch sector, some women have fished on boats with their partners and all of the oyster industry women in this research have worked manually on their oyster farms. Women's involvement in harvesting was the most significant difference between the case study sites. One possible explanation for this is that, in contrast to the wild catch sector, boats and wharves are not visible, masculine discursive sites of power in the oyster fishery. However, despite this apparently greater opportunity for women to engage in harvesting in the oyster fishery, particularly given that the oyster fishery has been established relatively recently and patrilineal inheritance is not as historically entrenched as in the wild catch sector, there is still evidence of a masculine industry discourse. Furthermore, in Cardea, no woman in this research had been fishing. This can be attributed to stricter cultural discourses of "wifing" and "mothering" for these women who are involved in ethnic fishing family businesses, as well as greater access to labour due to the prevalence of extended family involvement.

Women who engage in the practice of harvesting fish to assist their businesses in times of financial hardship are constructed as "helping". This is an acceptable activity which derives from discourses of "wifing" *providing* women in fishing make no claim to "fisherman" status. However, where women in fishing actively resist and attempt to claim the privileged masculine knowledge/ power of fisher(men) by fishing professionally as an occupation, they are re/constructed as abnormal and unnatural (as a "certain type", masculine, unclean and/or sexual) as punishment for breaching the discursive practices. Men in fishing and the majority of women in fishing enact this regulatory discursive mechanism. In addition, they police these breaches to support the dominant discourse and perpetuate the privileging of (fisher)men.

The dominant discourse constructs the collective of fishermen *as* the fishing industry *because they harvest fish.* Discursive regulatory practices reflect and perpetuate women's exclusion from fishing, support the relational and hierarchical dualistic construction of men in fishing ("fishermen") and women in fishing ("fishermen's wives"), define the industry as being comprised of individual fishermen, and exclude women from being visible or recognised as part of the industry. This results in the commonly accepted construction of industry stakeholders being fishermen and denies women any claim to stakeholder status, in turn perpetuating the construction that the fishing industry is comprised of and belongs to men who practice fishing.

> "Well, it's the **males who are out there catching. It's their industry"** (woman, rock lobster fishery, in fishing partnership, in her fifties).

In the following chapter, I further explore the masculine dominant discourse in industry decision making and how masculine industry ownership confers power to men in industry decision making and marginalises women in this arena.

CHAPTER FIVE

WOMEN IN INDUSTRY DECISION MAKING

Introduction

Having established that women in fishing lack visibility generally, this chapter explores women's visibility and contributions in a specific context-the critical sphere of decision making in Australian commercial fishing. It explores women's representation and positioning in industry decision making and the contributions women in fishing make to industry decision making, the barriers to and opportunities for women's greater participation, and the outcomes women would like to see as a result of this research. In this research, decision making in the Australian commercial fishing industry refers primarily refers to membership of and associated privileges in industry organisations, involvement in meetings, and committee and board leadership positions. However, it also acknowledges the less visible influence women may have in decision making. This chapter builds on previous research which has established that women in farming are marginalised in industry decision making (e.g. Alston 2000) and the small amount of Australian research which suggests that women in fishing may be marginalised in similar ways (Smith and Sullivan 2001).

I commence with a brief overview of women's representation in key industry organisations relevant to each case study. I then describe how the meanings associated with the subject positions of fishermen (stakeholder) and fishermen's wife (secondary) explain women's positioning in industry decision making. This is followed by discussion on the regulatory discursive mechanisms, including structural practices such as organisational processes and practices, which operate to exclude women in fishing and perpetuate their secondary positioning. Analysis is also provided on how some women in fishing resist these barriers and the consequences for these women who breach the dominant discourse. Strategies which may enhance the potential for the greater participation of women in fishing are then explored, including women's collective action. These strategies are viewed as both a form of resistance/ power (Foucault

1980) and as a mechanism for women's collective empowerment, and the basis for a standpoint for women in fishing (Harding 1987).

Women and Industry Decision Making

The fishing industry is characterised by a complexity of interrelated decision making structures, which makes it difficult to provide an accurate graphical representation of women's involvement in each case study. Some of these industry organisations are based on geographical areas, some on specific fisheries and others are based on specific aspects of fishing, such as processing and research and development. At a local level there may be fishermen's cooperatives and industry associations; at a state level, peak industry organisations; and at a government level, Management Advisory Committees (MACs) and Ministerial advisory committees. There are also decision making organisations, which have been established to address broader issues, for example, geographically specific environmental committees, with representation from various sectors, including government agencies and non government organisations, and representatives from the fishing industry.

At the time of this research, three state fisheries management agencies provided data on the representation of women in fishing in MACs, committees comprising of industry, community and government representatives to advise fisheries management agencies on management issues in specific fisheries. Representation of women in fishing on these committees was poor, with the highest estimate of women's involvement reported by these agencies being 5%, and the lowest being 1% or one in 101 positions.

While approximately half of the male partners of women involved in this research were involved at various levels in decision making from local associations through to industry boards, the majority of women were not involved in industry decision making. It is difficult to succinctly quantify the numbers of women involved in industry decision making given the complexity of decision making structures. Furthermore, it is difficult to provide figures which acknowledge women's contribution to decision making in less visible means, particularly where they are married to men who held committee or board positions. Furthermore, some women had previously been involved in decision making but had ceased their involvement, some women attended meetings irregularly and other women attended social functions after the conclusion of meetings where industry issues continued to be discussed.

However despite these difficulties, at the time of this research, no women in Cardea attended local meetings; three women of 17 in the oyster fishery held or had held positions on industry boards; and four women out of 34 in the rock lobster fishery had previously or currently held board or committee positions and/ or regularly attended meetings.

Western Australian Rock Lobster fishery

In the Western Australian Rock Lobster fishery at the time of this research there were 14 Professional Fishermen's Associations along the coast. Each Association president automatically gained a position on the peak fishery organisation, the Western Rock Lobster Council (WRLC). However, women were excluded from 12 of the 14 Professional Fishermen's Associations due to constitutional membership criteria. These exclusionary mechanisms, which meant that, at the time of this research, there were no women in fishing with full membership status in these Associations, will be further explained in a latter section, Organisations constitutions and practice. This fishery also had a Ministerial committee with 15 positions, the Rock Lobster Industry Advisory Committee (RLIAC) with significant representation from industry. At the time of this research, both the WRLC and RLIAC had 15 board positions and one designated position for female representation.

The RLIAC position was filled by a woman from a rock lobster family fishing business. However, while the designated WRLC position was technically being filled by a representative from a women's group, this woman was not actively participating or attending meetings. When discussing this, this women's group informed me that they were not actively pursing an association with this organisation as they has only been offered associated membership status.

Tasmanian oyster fishery

The research identified that, at that time, women were poorly represented as members of the Tasmanian peak industry organisation, the Tasmania Fishing Industry Council (TFIC). All wild catch license holders, fish processors and marine farmers – shellfish farmers including oyster farmers-automatically become members of this organisation. However, in the case of a company or partnership, only one person may be nominated for TFIC membership by the entity. A TFIC representative estimated this was a male 95% of the time. TFIC was not able to provide membership statistics by sex. However, given there are also licenses and leases held solely in the names of men and that only two women in Tasmania solely owned fishing industry licenses or leases, it can be estimated that TFIC's membership of approximately 900 individuals includes less than 45 women or 5% of total membership. A TFIC representative also estimated that from a total of 90 oyster businesses, there were only four women involved in decision making.

At the time of the research, there was one woman Director on the TFIC board, the first ever in a total of nine positions. However, the Tasmanian Agriculture Council, another peak organisation which is comprised of representatives in leadership positions from local aquaculture organisations, such as Tasmanian Marine Farmers and the Tasmania Shellfish Executive Council, had no women on the board. Three women in this research were attending meetings of smaller local oyster growers' organisations. At the time of this research, there were no women who held committee positions on either of these two associations.

Cardea – the Fishermen's Cooperative

The Cardea Fishermen's Cooperative, a fish distribution collective formed by European (fisher)men in the 1950s, was the main industry organisation in Cardea. This organisation provides representation and information to higher level decision making bodies relating to the trawling and longlining fisheries operating in this area. Members, who are also shareholders, are local fisher(men) who are currently fishing or who have retired. Shares are passed down through patrilineal inheritance in the names of individual (fisher)men, with no shares held independently by any woman in fishing, therefore excluding them from membership and associated privileges including leadership positions.

At the time of the research, the participation and representation of women in fishing in industry decision making is low. While there was diversity in the involvement and experiences of women in industry decision making and despite some women enjoying their experiences, the majority of women described decision making in industry organisations as "*male dominated*" with significant barriers to their attendance and participation.

Dominant Masculine Industry Discourse

The dominant masculine industry discourse privileges (fisher)men in industry decision making through their occupational and stakeholder status.

> "I don't know if I would actually become a member… **it is important that he take the lead** in this for the simple reason that **the professional fishermen make decisions** about what happens" … (woman, rock lobster fishery, in partnership, in her forties).

Deconstructing language reveals the industry *ownership* attributed to (fisher)men deriving from their occupational and stakeholder privileging. Local Rock Lobster industry organisations are called **Professional Fishermen's** Associations. In Cardea, the Cardea **Fishermen's** Cooperative, the local industry organisation that collectively makes decisions, has the following motto

> "The **Co-op belongs to the fishermen**, the **fishermen should belong** to the **Co-op".**

While men in fishing are constructed as "fishermen", women in fishing are constructed as secondary-*"wives" "behind their men"*. This construction is so pervasive that it prevails even when women do attend industry organisation forums that make decisions.

> "… very few [women attend meetings]. They tend to be **wives of fishermen**…or **wives of marine farmers, rather than attending in their own right"** (male representative, industry organisation).

This secondary construction of women in fishing as "wives", who lack the stakeholder status and thus the credibility to participate in industry decision making, was reflected and perpetuated through a range of masculine exclusionary practices prevalent in the culture of industry decision making.

> **"Culture**. It's pretty **hard with the bar open**…as far as **being accepted**. The **language**….I'm definitely there as **Pete's wife, not his partner**. The **men don't talk to me about fishing** … They will **talk to me about social business**…. if I thought I had a comment to make, I would make it, but … **it doesn't interest them**….It's very much a **bloke atmosphere"** (woman, rock lobster fishery, in partnership, divorced, in her forties).

These pervasive practices reflect the discourse that industry decision making is the "*place*" for (fisher)men/ industry stakeholders. In this "place", (fisher)men collectively display behaviours associated with the stereotypically masculine, "blokey" subject position of "fishermen". This behaviour reinforces the masculinity inherent in the subject position of "fishermen", and defines and perpetuates industry decision making as a masculine realm.

> "... the number of women ...who are actively involved... it's quite low....it has been a male dominated ... environment and it has a **reputation of being a fairly rough tough** kind of a **place** and people in industry have often **gained respect through being rough and tough**...I **don't know that it will ever fade,** it's one of the **unique things about the fishing industry**..." (male representative, fisheries management agency).

Women's place

Many women in fishing collude with the masculine dominant industry discourse that meetings-visible and public decision making forums-are not the "*place for them to be*" and subsequently have no involvement in industry decision making.

> "Some of the women ... and particularly the ones who I regard as being very capable of expressing themselves and quite knowledgeable, they have a view that it's **not a place for them to be**, meetings" (male representative, fisheries management agency).

However, several participants referred to women's *influence* in industry decision making. For example, some women do not attend meetings but *"tell their husbands what to do"* and *"who to vote for"*. This less visible involvement in industry decision making is acceptable as women do not claim stakeholder status. They continue to support the discursive positioning of women in fishing as secondary and the discursive formation of public decision making forums as "the place" of (fisher)men.

Many women in fishing also contribute to industry decision making in this secondary capacity by supporting or helping their partners. Women in European fishing families assist their partners with letter writing and interpretation. Eight women had also been "secretaries" in local industry organisations, despite lacking constitutional membership status. Several women were also "secretaries to the Secretaries", where they did the work of the position holder of Secretary in local organisations, which was held by their male partners. In comparison to the positions of Treasurer and President, who have responsibility for financial management and leadership, the traditionally feminised position of Secretary is also relatively powerless.

> **"...being the Secretary,** they just went **'Well, that's what you are. Just write down what we want you to write'"** (woman, rock lobster fisher, in partnership, divorced, in her forties).

This involvement draws from broader "wifing" discourses that position women in fishing as doing *"what needs to be done"* to assist their partners and preserve the family economic unit.

> "...keep **building on what we have got to get in a good position.... my husband's quite involved** on committees ...so **I help him write letters"** (woman, rock lobster fishery, in partnership, in her fifties).

These activities are normalised through the regulatory discursive mechanism that renders the skills and qualities required for these activities – attention to detail including a proficiency in paperwork, organisational capacity and being generally supportive-as "naturally" embodied in women.

The discursive place of wives and their partners

The construction of women in fishing as "wife" and secondary to their partners means that women's level of participation in industry organisations is influenced by and, in some cases, *determined* by their partners. Thus, many women do not participate in industry decision making *because* their partners are not interested or involved.

> **"Paul is not a member** of any ... fishing organisation, so, no, **we don't go** to the meetings" (woman, rock lobster fishery, in partnership, in her forties).

Other women participate when they are "*needed*", when they are "*encouraged*" or "*allowed*", or when their partners *"ask"* them to become involved.

> "...**different people have asked me** to get involved and **I said 'No** no'. And **Sam [husband]... said 'When are you going to get involved?'** and 'Oh, they don't want **outspoken people'**, and so they said 'No, **we need**

> **you'"** (woman, rock lobster fishery, in industry decision making, in her forties).

Receiving this "permission" provides women with an understanding that their involvement does not constitute a discursive breach which potentially could threaten their personal relationship and family economic unit. Receiving this encouragement from their partners resulted in several women becoming empowered and facilitated their further involvement in industry organisations. However, many women in fishing recognised that participating *against* the wishes of their partners constituted a significant discursive breach with potential significant implications for their personal relationship and family economic units.

> "They'd probably **get divorced**….**some men don't like it**…. in the end the **men want to make the decisions"** (woman, rock lobster fishery, in partnership, European, previously in industry decision making, in her sixties).

Barriers and the discursive wife

Women in fishing face many barriers to enter into and participate in industry decision making. Many of these barriers are masculine, exclusionary practices prevalent in industry organisations and fisheries management agency decision making forums that function to keep women "in their place". However, there were also many barriers for women in fishing relating to their daily activities such as a lack of time, family and childcare responsibilities, unpaid fishing business commitments, time away from family and the business, cost of travel, paid work commitments, cost and lack of childcare, geographic and social isolation, increasing debt and a lack of qualifications and board experience.

These barriers reflect acceptable "natural" activities for women that derive from broader discourses of "wifing" and "mothering" and are inferred through being a *"woman"* or *"female"*.

> "I can't see the day when the husbands say 'You go to the meeting, I've got to pick the kids up from school'…. **It's the traditional role of the female…she still has the role of the house and the children and those will always come first with her**" (woman, rock lobster fishery, in partnership, in her thirties).

The normalising function of discourse, which results in many women in fishing adopting/ being relegated to activities deemed "natural" and acceptable for women, supports a *discursive* division of labour where "fishermen's work" is characterised by fishing/ sea/ public/ visible/ decision making and "fishermen's wives work" by family/ home/ private/ invisible/ non decision making.

> "No, I never ever wanted to [get involved]... I **just left that entirely up to my husband**, purely because **that was his work**, and this was **my work here"** (woman, Cardea, in her fifties).

The conceptualisation of the division of labour as *discursive* reflects the potential for women's agency to adopt or resist engaging in activities associated with differing subject positions. This is in contrast to the division of labour being employed as a functional, essentialist and tautological explanation of the different roles and work activities associated with the categories of "women" and "men".

One woman deconstructs the poor representation of women in fishing in industry decision making to expose the discursive division of labour as the *culture* within the industry.

> "...[the lack of] training, or crèches ...I really don't think it's that... **it's more the culture** in the fishing industry. The **women** ... tend to **stay home and look after the children** while the **husband's at sea** and **do the bookwork"** (woman, key informant, in industry decision making).

Women's diversity

Diversity within the range of subjectivities of women in fishing influenced their capacity to become involved in industry decision making. Women who were single parents, younger, disabled, of European background, or from a rural area were further marginalised in industry decision making. One woman in fishing who operated her own oyster farm spoke about how being a sole parent intensified economic and family responsibilities and made it impossible for her to commit to a position in an industry organisation. Women in fishing who lived in rural areas were faced with greater costs and distances to travel to attend meetings and commonly experienced difficulties arranging childcare. One woman also described how disability had barred her from attending meetings and rendered her less visible.

Women who married European (fisher)men were bound more strictly to acceptable (silent) activities for women through a stronger cultural discourse of "wifing" and "mothering". One woman in fishing explained that in this discourse women "should be seen and not heard". Another woman described racial discrimination in industry organisations as an additional barrier to her involvement.

> "…**they think because you come from another country you're dumb** … you're not good enough. They…**talk to me so slow** because they think I don't understand English….because I'm different…. **they don't know how to talk to you, so they just ignore you. They don't talk to you at all"** (woman, oyster fishery, in partnership, ethnic, in her forties).

The Way the Industry Keeps Women Out

Exclusionary discursive practices that privilege (fisher)men with occupational and stakeholder status, function to preserve industry decision making as the arena of (fisher)men. In many industry organisations women are excluded through constitutional membership criteria, nomination and voting practices, selection criteria and processes and through a dominant masculine culture. This includes the timing and location of meetings and training, masculine behaviours that reinforce and privilege the subject position of "fisherman" and women's lack of "innate" fishing knowledge (power). In some industry organisations, gender diversity is being addressed. However, the dominant masculine discourse remains prevalent in these organisations, with women's exclusion perpetuated and justified through discourses of merit and fairness, and the view expressed by some men and women in fishing that there are no barriers to women's participation.

Organisational constitutions and practices

Women in fishing are excluded from participating in industry decision making in several industry organisations through constitutional membership criteria which privilege the ownership of fishing capital, an area where women are marginalised. In the Cardea Fishermen's Cooperative, the transference of shares through patrilineal inheritance meant that no women were members. Consequently, women were not entitled to attend meetings or access information such as financial statements. In other industry organisations, women without sole ownership of a boat or license were not able to vote at meetings other than as proxy for their partners.

However, in some cases, women in fishing are marginalised *irrespective* of ownership status. In the rock lobster fishery there were two main types of licenses. A Professional Fisherman's License costs approximately $100 and is an essential requirement for all skippers and deckhands – (fisher)men-to commercially harvest fish from a boat. This license is distinct from the Managed Fisheries License, the license required by businesses to operate in the industry. These businesses are worth up to several million dollars. At the time of this research, membership eligibility in 12 of the 14 Professional Fishermen's Associations was contingent on the applicant having a Professional Fishing License. This meant that deckhands could join with full membership, and access voting and nomination privileges, without having any other industry interest. However, women in fishing, regardless of their sole or partial business ownership status are not eligible to join as full members unless they also have a Professional Fishing License. While technically these women could apply for a Professional Fishing License, the majority of women do not have this license due to their lack of participation in (exclusion from) the practice of harvesting.

The marginalised membership status of women in fishing excludes them from involvement in industry organisation nomination and voting processes. In TFIC, nomination forms were sent to members every two years to vote for Directors. Women involved in entities where they are *not* the nominated business member, estimated by TFIC to be 95% of family fishing businesses, were not able to stand for nomination or vote independently. While postal voting may be seen as a strategy that facilitates women's input, women in fishing remained marginalised where there was only one vote per business, as fisher(man) status privileges membership.

In some cases, women's restricted capacity for nomination meant they were further excluded and not able to represent local organisation on peak organisations. In the rock lobster fishery, Presidents of Professional Fishermen's Associations automatically held positions on the State's rock lobster peak organisation, the Rock Lobster Industry Council. However, women's membership ineligibility in the majority of Professional Fishermen's Associations meant they were unable to hold the position of President, which is a necessary requirement for representation on the peak organisation.

The pervasiveness of the dominant masculine industry discourse in decision making is reflected in many women being *unaware* they were unable to become members of some organisations. While some women collude with the dominant masculine industry discourse by expressing their view that their membership status is not important because their partner makes decisions or because they always agree with their partner, other women resist the lack of stakeholder status of women in fishing.

> "…**I'm not allowed to vote**…and … **hold office because I'm a woman and their constitutions say that I can't**…. **the fact that the association incorporated document says that a female can't hold office really pisses me off**….I've got George saying to me 'Come and be the President of the association'….But I can't anyway because it's just **bizarre that in 2003 there's a body of people with an incorporated document saying that women don't have a part"** (woman, rock lobster fishery, in industry decision making, in partnership, three children, in her forties).

> "You'd probably be an **associate member even if you were half a bloody license owner**. They think you are a **second rate citizen"** (woman, rock lobster fishery, previously in industry decision making, European, in her sixties).

These practices operate to exclude women in fishing and preserve power for those in industry organisations-(fisher)men.

> "… you should be able to [vote]…. why should I have to sit in the back stalls?....I just think that **they are trying to keep us out** of it for as long as they can…" (woman, rock lobster fishery, in partnership, in her fifties).

Despite these exclusionary practices, there is some recognition of women's poor representation in industry decision making. In RLIAC, at the time of this research, vacancies were advertised in newspapers and through fishing industry organisations and members appointed by the fisheries Minister. Yet, while this organisation was trying to increase gender diversity on their board, women in fishing were not targeted specifically in these advertisements.

Two of the 14 Professional Fishing Associations had changed their constitutions prior to fieldwork to enable women in fishing to join as full members without a Professional Fishing License. As a result, women's representation at some meetings increased. However, given the pervasive masculine industry discourse, significantly increasing women's representation through constitutional change alone is an unlikely outcome for many organisations. One male representative of a Professional Fishermen's Association that had recently changed it's constitution to enable women to become members, felt that this change was a sufficient strategy to increase women's representation. He indicated in a surprised and joking manner that "the guys" hadn't considered further strategies to facilitate women's increased participation.

> "…the **first step is** opening up the Professional Fishermen's Association **to women membership** … Um, but no, **we don't have a strategy** …. I suppose we'd have to look at it. Don't think **the guys have really thought about it.** But now you mention it, maybe we should!" (laughs) (male representative, rock lobster industry organisation).

A further issue is that this constitutional change was not based on social justice principles which acknowledge women in fishing as industry stakeholders. Instead, in a climate of industry insecurity, *"getting the wives involved"* is a mechanism to increase the organisations' numbers (presumably for political gain) and enable women *to do the work.*

> "Oh well, the **numbers are dropping off**…. So we thought, one way to do it is to **get the wives involved**….it would **make our job easy**…take **half the workload**…. fishermen…just want to go and fish and leave the rest of it to everyone else" (male representative, rock lobster industry organisation).

This pervasive secondary positioning of women in fishing and their lack of stakeholder status remains an exclusionary barrier for women in industry decision making and perpetuates the discourse where women-"fishermen's wives" provide support and help when required and "(fisher)men" remain privileged with occupational and stakeholder status.

"Fairness", "merit" and "no barriers"

In one state peak industry organisation where women were marginalised in accessing membership status, a representative expressed his view that voting and nomination processes were *"fair"* and that gender diversity at board level was a non issue.

> **"I don't think we have any concerns about it** [gender equity on the board] if you understand the process, which is that every member gets a ballot paper and …can nominate to be a director…. and it therefore means that **everyone has a fair opportunity to nominate, a fair opportunity to vote**…" (male representative, industry organisation).

A representative from another industry organisation, where women were also marginalised through membership status, described similar "*democratic*" voting procedures. This gender blind discourse of "*fairness*" is synonymous with that of "*merit*" which perpetuates existing practices that privilege men and *disguises* gender inequity and women's exclusion.

Several participants, including half the women involved in industry decision making and male representatives of industry organisations and fisheries management agencies, also believed there were "*no barriers*" for women to participate as decision makers. This view being held by women who are involved can be explained by their desire to protect the privileged status they have fought to gain and can be seen as a mechanism to keep other women "in their place". This collusion with the dominant masculine industry discourse perpetuates industry decision making as a masculine domain by attributing women's poor participation *to women,* without attributing any responsibility to those in power-typically men-to alter the masculine culture and structural practices that exclude women. The fallacy of "no barriers for women" was also revealed as many women in industry decision making with this view made contradictory statements indicating there were barriers for women.

> "… absolutely **nothing to stop them** …there's **no barriers** …. **As long as she has the support of her partner or husband"** (woman, key informant, in industry decision making, in her fifties).

Culture

Many women in fishing describe the masculine exclusionary practices of a "*men's only club*" characterised by lengthy board tenures, selection processes based on an "*old boys*" network-the "*right school*", "*right tie*", and the "*right family background*", a double standard for men and women, a "*concrete ceiling*" and being perceived as the "*token*" woman. The contributions of women in fishing and their stakeholder status are not recognised and women's exclusion is reflected and perpetuated in the pervasive construction of women in fishing as "fisherman's wife".

> "If you go along to the meetings and hear the men talking…it's a **man's world**….I'm **the only wife**….**They don't really recognise women**, they **don't really recognise my role** in the business" (woman, oyster fishery, in partnership, in industry decision making).

> "... **[women] probably feel intimidated by the men**. Like if they say the wrong thing it's like you don't know anything. **'You're only a wife! What do you know?"** (laughs) (woman, Cardea, in her fifties).

Women in fishing are excluded and marginalised through the masculine culture of decision making. Some organisations hold meetings in pubs and clubs. Many women describe being ridiculed and not listened to; meetings characterised by drinking, shouting, *"table thumping"*, swearing, sexist jokes and instances of sexual harassment; and men not discussing industry issues with women.

> "...there's still a fair bit of **macho-ness**...if there's a board meeting, then they all go to the **club** and ... **drink** and there's that **male bonding thing** ....[women are] **excluded** from ....they're **not privy to information**... during those...**casino sessions"** (woman in fishing, key informant, in industry decision making, in her forties).

While some women attend social functions held *after* meetings, typically women are *not invited* to industry organisation meetings, conferences and overseas tours.

> "I've never been to one of their meetings.... they **invite blokes**... it's **down at the Club** and they **have beers** and it's like a **male club"** (woman, rock lobster fishery, in partnership, in her forties).

Women in fishing are also typically not invited to industry organisation pre and post season gatherings which are also often held in the pub and characterised by heavy drinking. In some cases, strippers are engaged for these events. One woman recognised that if she attended, her partner would be subject to a penalty by other (fisher)men for breaching practices that defined these events as masculine. As a consequence she did not attend.

> "If they had a pre season drinks ... **it wasn't actually said 'You are not invited', but we all knew we were not wanted.** You don't go along **because [how] it would look to the other guys if they brought you along**?....and then [I] said to my husband 'Well, who would really want to go to your **drunkard barbeques on the wharf**? Not me" (woman, rock lobster fishery, in partnership, in her fifties).

Women's lack of stakeholder status in the fishing industry is reflected in the inclusion of women in fishing as *arbitrary* by some organisations. In recent years, women's resistance had resulted in their inclusion in post season celebrations. However, one woman also described how, after a financially poor season, one local organisation reverted back to a more "traditional" masculine (exclusionary) celebration, perpetuating the masculine privileging in industry organisations.

> "…end of year wind ups! …. **if the industry, if men were to realise… it's both the husband and wife that was playing a role… they'd be having something that the women and the men could go together to**….years ago they used to have … **piss ups** … **with strippers** … but the **women kicked up a bit of a stink** and said 'Hey! We want to have a time when we can get dressed up and have end of year dos'…so they did, but then… they said because **we don't have very much money this year**, we're going to have to can it….[And **they did] the same thing they used to** do! It's just **an excuse** to revert back to stupidity….**that's just one way I can say the … industry doesn't view women like they have any role within it"** (woman, rock lobster fishery, three children, in her twenties).

Knowledge (power)

Many participants expressed the belief that women in fishing lacked sufficient knowledge (power) to participate in industry decision making because they did not harvest fish. The exclusionary nature and pervasiveness of this dominant masculine industry discourse was revealed in comments by male representatives of industry organisations.

> "Basically the trawlers are all getting together …to fight the government and ….we had a meeting here just the last few weeks ago…."
> Jane: "So did any of their wives come?"
> "Oh no….no **(laughs)**. Well **they don't know enough about it to put an opinion in"** (male representative Cardea, industry organisation).

Women are excluded from industry decision making on the grounds they lack knowledge and practical fishing experience.

> Participant 1*:* "If you **dared stand up** and **said your opinion** …"
>
> Participant 2: "'**Shoot her down"**
>
> Participant 1: "It's like **'sigh, here we go, it's a woman'"** (laughter)
>
> Participant 4: **"'What would she know? She's not out on the ocean'"**
>
> Participant 1: **"'How could she possibly know anything?'"** (women, rock lobster fishery, focus group)

> Participant 4: "…[women's] opinions are not really valued…**the fishermen think 'Well we don't catch fish, so we don't really know what is going on'**… it's just a **male dominated industry"** (woman, rock lobster fishery, sole owner, in her forties).

Language and meaning in the following comment by a male industry organisation representative reveals the potency of men's industry *ownership* and privileging ("*their*" case and "*their*" interests) which is conferred through occupational status-*"the bloke"* does *"all the hard work"* and they *"know"*. This masculine privileging reinforces decision making as the arena of (fisher)*men*.

> "I don't think there are very many women who get involved in the direct decision making…. it tends to be the **bloke that goes along** and does the arguing….these are **the blokes that do all the hard work, they know all the practical issues… they're the people that have to argue their case and look after their interests"** (male representative, oyster industry organisation).

Several women in fishing resist this discourse by *valuing difference* and recognising that despite women's lack of fishing experience, they still have "knowledge" (power) and a contribution to make.

> "…women do have a significant impact on their family business … and **I don't think they're given enough credibility to what their knowledge is** and they probably don't allow for that in forums…" (woman, rock lobster fishery, in partnership, in her fifties).

The fallacy of women's lack of knowledge also became evident during interviews as the majority of women, irrespective of whether they do or do not claim knowledge, are able to comprehensively discuss industry issues.

However, this alleged lack of knowledge of women in fishing is a particularly pervasive barrier for women given the construction of fisher(men) embodying knowledge (power) through their alleged *"natural"* connection with fishing. The masculine discursive practices that exclude women from gaining experience in fishing also exclude women from accessing this knowledge (power). Therefore, this knowledge (power) is inaccessible to the majority of women in fishing. In turn, women's lack of knowledge (power) justifies their exclusion from industry decision making and women in fishing's poor participation reinforces the commonly expressed belief that women *"do not"* (cannot) *"know"*.

Masculine exclusionary practices also may be seen to result in and explain women's lack of confidence which is one of the most commonly reported barriers to women's participation in industry decision making.

> "... it's confidence. They do have the ability to do it, but they are quite **scared** of...**whether they are actually going to be heard** and **whether they have something to contribute** and **whether people will actually listen** to them **and respect their opinions**..." (woman in fishing, key informant, in industry decision making).

Women's lack of confidence may be further deconstructed to reveal an additional regulatory discursive mechanism pertaining to women's exclusion from industry decision making. While women's poor representation in industry organisations is attributed to women's *own* lack of confidence – that is, where *women are* constructed as *responsible* for their lack of confidence-there is no responsibility on men to alter existing exclusionary practices, which at least partially *account* for women's lack of confidence. However, recognition by some women that women in fishing do have the capacity to participate, suggests the potential to empower women and increase their representation in industry decision making by exposure to alternate discourses that recognise the knowledge, skills and capacity of women in fishing.

Women's Resistances

Women in industry decision making

Despite the prevalence of exclusionary practices, some women in fishing resist and participate in industry decision making. Women's experiences in industry decision making and leadership positions in industry organisations are diverse. Women influenced by exposure to alternate discourses of gender and equality through education, family background or work experience are more likely to participate in industry decision making and less likely to adopt traditional gender roles and values. This exposure resulted in these women having the *expectation* of being accepted, respected and acknowledged as industry stakeholders.

> "I have **a family history of achieving women**.... the expectation would be to get out there and do it.... and **a father who just expected the same from his daughters as from his sons**....I have always...**expected to be accepted"** (woman, oyster fishery, not in partnership, tertiary educated, in her fifties).

"...**I'm not a silent partner**...Now they are realising I'm **not just a pretty face. I'm part of the industry.** I do the work as much as they do .... I'm certainly **made to feel welcome now that they realise what I am....and they'll talk oyster talk with me, rather than just female talk"** (woman, oyster fishery, in partnership, second marriage, in her sixties).

Other women in fishing become involved in industry decision making where there is a perceived threat to their family economic units. Two women in European fishing families attended meetings to assist with the language barrier that marginalised their fishing communities when dealing with fisheries management agencies. These women resisted industry racism and resistance from their partners for attending in order to benefit their family economic units and local ethnic fishing communities.

"...my husband's relatives ...when they're trying to communicate with 27 year old economic students...there is a big language barrier ....**every time we have a meeting, I'm writing a letter** to the government representatives saying ...if our guys could speak in Italian ... they would get the message across....and because they can't ... it's a miscommunication...**because of our names people automatically assume an Italian immigrant....'No speaka de English'. 'Excuse me! I'm fourth generation Australian.**..I've got blonde hair and blue eyes'....Totally different attitude" (woman, Cardea, in her fifties).

Women involved in industry decision making also have to *manage* the masculine culture. Some women feel that women's participation is only possible if women accept *("cop")* practices associated with this masculine culture.

"...it's a **cultural thing** about the fishing industry being **hunters and gatherers** ... being a **blokey thing** and I have certainly found **no impediments** to what I've been doing at all, but **then you have to cop a fair bit of** ... **bad language and sexist jokes** ... I mean **that's just part of the industry**....that comes back to your **blokey culture**....There's **not much political correctness**..." (woman in fishing, key informant, in industry decision making).

Other women in industry decision making draw from discourses of natural gender differences and employ "feminine" strategies to assist them in managing the masculine culture.

"....**they're not game enough to have a dig at me because I might get offended and start to cry** ..." (laughs) (woman, rock lobster fishery, in partnership, in decision making, in her forties).

Despite having to manage this masculine culture, some women involved in industry decision making enjoy participating and become empowered. One woman shared this alternate discourse with her daughters, despite the pressures in managing conflicting aspects of identity within her subjectivity such as "mother"-such as feelings of guilt because she was not always available to her children-and empowerment gained through being acknowledged and participating in industry decision making.

> "…my kids…[have] said 'oh, but … **you're going away again**!'…and I've … said 'Well, **what do you want me to do? Do you want me to just be a mum that stays at home?...**'my eldest daughter…said '…I was just so proud of you Mum, cause you've done all of these things!....You're **such a good role model'**, and I thought, **Well, the little ones might not think it yet**, but I think eventually **they will realise that if all I can teach them is that if you believe in something bad enough do something about it, then that would be fine"** (woman, rock lobster fishery, in industry decision making, four children, in her forties).

Expression as resistance

If the expression of alternate discourses reflects empowerment and may be considered as a precursor to political action, then women in fishing who express support for alternate discourses may also be seen as resisting. Some women claim that women in fishing have a *greater* knowledge than (fisher)men.

> "…**some women … know more than what most fishermen do**. I've been to … meetings **and you sit there and think 'What are you talking about?' Even I understand it better than you, and they're fishermen!"** (woman, rock lobster fishery, not active in decision making, in her fifties).

Other women in fishing feel that (fisher)men's lack of experience outside the fishing industry, resulting from the generational transfer of businesses, means there are significant gaps in (fisher)men's knowledge (power). These women employ an *expanded, more holistic* construction of the fishing industry as *more than the harvesting*, and resist masculine industry privileging through this re/construction.

> "We need to understand more about the industry on a world basis…and I believe that **a lot of fishermen don't understand that,** the exporting and marketing demand…. [it's] a generational thing. They get off from school

and jump in a boat.... they're in their little small cocoon" (woman, rock lobster fishery, in partnership, tertiary educated, in her thirties).

Some women demonstrate further resistance by developing this expanded construction and claiming stakeholder status for women, families and communities.

> "The fishing industry is a very male focused industry...Women have not been recognised. They've been the **invisible partners** whether they work on the boat or whether they work at home.... there are **more people in the industry than just a bunch of blokes in boats... we're families... we're communities...**" (woman in fishing, key informant, visible in industry).

Resistance against women's lack of industry stakeholder status is also evident. Employing feminist research principles, I responded to questions by four women on my understanding of their eligibility to join their local industry organisations. This exposure to subjugated knowledge generated passionate responses from each of these women. This indicates the potential power to alter subjectivities through exposure to alternate discourses which reveal the gendered relationship between financial investment and stakeholder status.

> "....**when you are financially involved like that there is no reason why you shouldn't have a vote!** It seems to me that you are very much an **industry stakeholder"** (woman, rock lobster fishery, in partnership, in her fifties).

Resistances Against Women who Move into Leadership

Women in fishing who participate in industry decision making are derogatorily re/constructed by some men and women in fishing for breaching the masculine privileging in industry decision making. The positioning of women in fishing as secondary is evident in the persistent construction of women in industry decision making as "*wives*". Women in industry decision making are also perceived as lacking embodied fisher(men)'s knowledge (power) and are not able to participate without being perceived as "lesser" (than men), *irrespective* of the strategies they employed to manage the masculine culture. When women in fishing adopt masculine practices and behaviours, they are derogatorily re/constructed as "*being like a man*". When women resist and do not adopt masculine practices and behaviours, they are derogatorily re/constructed as "feminine" and "weak" and not being "tough enough" to participate. Many women in fishing collude with the dominant masculine industry discourse and derogatorily re/construct women involved in industry decision making as either "like a man" or as bad and selfish "wives" and "mothers".

"Fishermen's wives" in industry decision making

Women who actively participated in industry decision making experience resistance by persistently being constructed as "wives".

> "I walked in the door and **they said 'Ok, so this is a housewife'**. I **don't have a degree** behind my name. I **don't work in a department**… I'm **not associated with a university**. Effectively **I'm not even a fisherman. I'm just a housewife**. So and it was really, **really difficult** to sort of walk in there …'Oh shit' … then you **have to have the confidence to say 'No, you're wrong'** "(woman, rock lobster fishery, in decision making, in her forties).

Resistances against women in fishing who actively and publicly participate in industry decision making stand in contrast to women who do not challenge their positioning as secondary and as "wife" by attending *with* (supporting) their husbands and remaining *silent* (not speaking).

> "I went to a couple of …[meetings] …**I was young** and **nobody wanted to listen** to what I had to say .. **I felt like such an idiot**…I was **frowned upon to have an opinion**…there was another woman there…and she was a lot older…She **didn't say anything, she was just there with her husband"** (woman, rock lobster fishery, sole license owner, in her twenties)

Several women involved in industry decision making, while resisting this construction of themselves, collude by reinforcing the constructing of other women in fishing, who are not involved in industry decision making, as "wives".

> "I don't know what it [the experience of participating] would be like **if you were a wife"** (woman, oyster fishery, in decision making, tertiary educated, in her fifties).

At *all levels* of industry decision making, collusion by men is evident in discursive practices that function to reinforce the status of women as "wives".

"….particularly at a **higher level, at both state and federal level,** it is a **boy's club**….those sorts of boards are **very good at asking women to make the cups of tea.** So we **say 'I'm sorry, I don't do tea, but I do sit at the table'"** (woman in fishing, key informant, in industry decision making).

Women in industry decision making are punished for discursive breaches. One woman described her experience of social ostracism by her extended family and broader fishing community.

(Laughs)" **It's not the done thing. Excuse me! I'm supposed to be home looking after my husband, cooking his meals and making him chocolate cakes.** I'm a little bit of an **enigma**, some of his relatives don't quite know how to handle me**. Some of them have actually stopped coming [over] because I'm too outspoken….to go to a meeting cause there's all those men there!"** (woman, Cardea, in her fifties).

Women in industry decision making, who participate against the wishes of their partners, also experience resistance from these partners, with several women describing instances where this had resulted in divorce. While one woman from a European fishing family experiences hostility from her husband, her participation in decision making forums with representatives from both industry and fisheries management agencies is tolerated due to his *need* to have his wife assist in resisting fisheries management agencies.

"[He says] **'You're not supposed to go there and talk to all those men!'** (laughs) and I turn round to him and say '**Well you go**!' and he goes '**No no no no… Ok you go'**…**I'm supposed** to do **what his mother does** but I'm also supposed to do **what his father does"** (woman, Cardea, in her fifties).

Another woman, in the rock lobster fishery, had rigidly adhered to the traditional role of "wife and mother" until her forties when she became involved in industry decision making. This woman resists pressure that she remain a "housewife" and is punished through passive aggression from her partner. The substantial level of conflict is indicated in her threatening to leave her marriage.

"He hates them [decision making commitments]. Absolutely hates them. And **every time I go to Perth I have to put up with 2 days silence**…. I said 'That's it. **I've had enough. You either let me go to work or you let me go to school or I'm out of here' because I cannot be a**

housewife....It's just not me....So he's learning to deal with it" (laughs) (woman, rock lobster fishery, in industry decision making, in her forties).

Further comments from this woman reveal the relational and hierarchical constructions of the subject positions of "woman" as traditional wife and mother and "man" as the more powerful male "breadwinner". The acknowledgement this woman receives for her contributions in the public decision making arena undermines her "man's" power. This appropriation is reflected in the language of "*gone from being "his wife" to "he's my husband"*. In order for her to continue her activities, this woman also has the responsibility of maintaining stability of the family economic unit in the face of "*a huge amount of aggression at home"*. This is the cost of her discursive breach, where she has resisted her secondary positioning as "fisherman's wife".

> "Personal cost?....a **huge amount of aggression at home**...he doesn't cope at all... **I have to fight every single time** I want to do something ....and a lot of other women would have to do the same thing...It's a **big thing for men**, if their wives haven't worked...for my husband who's gone from being the breadwinner, to the person whose now 'Patty's husband'....**I've gone from being 'his wife' to he's 'my husband'** And very difficult for a man's pride..." (woman, rock lobster fishery, in industry decision making, in her forties).

Women who move into industry decision making have to manage the conflicting aspects of their subjectivities in terms of the expectations and requirements of being involved in decision making, and the pervasiveness of regulatory discursive scripts. These scripts relegate women to/ encourage them to take up "natural" secondary activities associated with the subject position of "fisherman's wife" and traditional wife and mother roles.

> "...when you come home from a conference, **nobody....says...'Tell me about your weekend'... Nobody cares**... **You come home with this massive amount of importance** because people have ... thought you were intelligent ...and **all of a sudden ... you're just the finder of lost socks** ...and **it's lunches** tomorrow and **school uniforms** and **'Mum, I need five bucks for this'"** (woman in fishing, key informant, in industry decision making).

Masculine discursive practices as resistance

While masculine discursive practices in industry decision making, such as "*table thumping*", swearing, "*cracking open a tinnie at the end of the meeting*" and "*shouting each other down*", may be seen as exclusionary practices preventing women in fishing from becoming involved, they can also be seen as *resistances* against women who do participate.

> "…if they [women] go there the first time and get bagged or don't get listened to …. **Blokes are bloody notorious**. A lot of them talk over the top of a woman … if it's their first foray into the arena…they go alright 'I'll get back into my shell'...It's only stupid idiots like me that keep coming back…**You get a thick skin** on you. There are times where you think '…**How much more of this rubbish do you have to take?'"** (woman, rock lobster fishery, in industry decision making, in her fifties).

While many women describe meetings as "male dominated" or a "man's world", some women recognise that some men with "less masculine" subjectivities also find this behaviour affronting.

> "…there are one or two people who need to **clean their act up** a bit and just **disrupt the meeting**…who have got **foul mouths and senses of humour** and **that to a woman is pretty off putting**. They don't want to be there, and **even some of the guys are embarrassed**…" (woman, rock lobster fishery, in industry decision making, in her forties).

Women require strategies to *manage* their participation in this environment. However, *irrespective* of the particular strategies they employed, women are derogatorily re/constructed. When women do not adopt requisite masculine behaviours and practices, they are constructed as weaker than men and as lacking the capacity to participate in industry decision making.

> "...the director…was a table thumper, so were a whole lot of the other ones, and they just didn't give a shit about consensus! It was more the **bulldozer approach … a male approach** … **she didn't come across like that** and **that's often seen by men who are table thumpers as a real weakness"** (woman in fishing, key informant, in industry decision making).

However, women in industry decision making are in a no-win situation as women who *conform to and adopt* prevalent practices and behaviors, are derogatorily re/constructed as being "*a man*" and as lacking intelligence for displaying the masculine qualities associated with the subject position of "fishermen".

> "...women in leadership situations **have to be a little more confident** and more **boisterous** which is sometimes **viewed as 'She'd just a mouth woman with not much intelligence'"** (woman oyster fishery, in partnership, in her forties).

The majority of women in fishing who did not participate in industry decision making also derogatorily re/construct women in industry decision making as masculine and lacking intelligence. One focus group punished a woman in industry decision making harshly for her involvement as giving "*us all a bad name*". This was a standard which was not applied to any men during the course of the research and this woman no longer participated in industry decision making.

> Participant 1: "We **used to have** a lady, Grace... she was **a joke**, and she was **dreadful** and she **really gave us all a bad name** because she'd stand up and make such a **fool of herself. She hasn't been for a while"**
>
> Participant 2: **"No, she has improved...."**
>
> Participant 3: 'Because she was **one of the first ones to start to speak her mind**, they used to egg her on and that. **Unfortunately** when a lot of women do branch out, instead of being women**, they try to be one of the boys**..." (women, rock lobster fishery, focus group)

However, *in order to* enter into and compete in industry decision making, women often have to not only accept, but also *adopt* these masculine practices.

> "There's a couple of women .... Their husband's had actually died.... They were really **abusive** to her. They got really **nasty** ... they actually **yelled her down....I admit she was aggressive to start with but maybe she was that way because she has experienced this war that there is. She would have to be like that just to get heard"** (woman, rock lobster fishery, in her forties).

Despite her own personal negative experience of participating in industry decision making, one woman derogatorily re/constructs other women who are involved for challenging the status quo by employing her construct of "equality" where "*doing the equal thing*" means that women participate *in the same way* as men participate. This construction punishes women for their "difference" to men and strongly colludes with the dominant discourse by maintaining the masculine culture in industry decision making.

> "I've seen some women and they're **atrocious!...**and I've seen women do it **[breastfeed] at board meetings**, because **they say that it's their right**! If we wanted to be treated equally, **I haven't seen any men bring their toddlers** to a board meeting....We had a [meeting] and a woman rang up and said, 'And what babysitting requirements have you made for the meeting?' and I said 'None, seeing the meeting starts at 8 o'clock on Monday night'.
>
> She said 'You're denying me the right to come'.
>
> I said, 'Well madam, **find your partner if he wants to come**, but at 8 o'clock at night I don't feel that I have the responsibility to provide childcare facilities'.
>
> And she said 'Well you're creating a barrier for me coming'.
>
> And I said, 'It's your **choice**....'
>
> **Who takes a child to a public meeting at 8 o'clock at night?** I mean in all honesty!....**Why don't men bring their kids to meetings?....**I just think that **women trade on it, and they want equality but they don't want to do the equal thing"** (woman in fishing, key informant, in industry decision making).

This woman also punishes women in industry decision making for resisting sexual harassment. The construction of women as being overtly sexual (*"playing games", "sexy skirts"* etc) constructs women as sexually promiscuous and as *responsible for* the sexuality in some industry decision making forums, thus absolving men of any responsibility to alter this culture.

> "...the sexual harassment bit! The women turn up in their **low shirts, low tops and their little sexy skirts, score all the boys** at all the conferences and then **yell out 'They have no respect for me'**. You **can't play those games. You can't play 'I'm a woman, I've got tits, and I'm really cute and gorgeous but don't look at me'**....A lot of the men also get really scared because they're not too sure what's going to be sexual....**Where does that leave ... those ... 50, 60 year old men...that have been calling everybody 'love' and 'sweetie' all their lives?**....they don't mean it as that....**and these younger women who are coming through, 'That's a sexual come on!'** No it's not. He's happy to see you "(woman in fishing, key informant, in industry decision making).

Many women in fishing also collude by asserting that women lack the knowledge and practical experience. As women are excluded, (fisher)men's "knowledge" is synonymous with the credibility required to participate.

> **"I don't think** that there are **enough women in the industry that have had the experience to make the decisions**..." (woman, oyster fishery, retired, in her sixties).

Some women in fishing collude more rigidly with the masculine privileging of industry decision making and espouse the view that "*women*" do not have equal capacity to know purely *because they are women, regardless* of their practical experience.

> "... there's **even women fishermen that go out cray fishing** ... Good on them if that's what they want to do...**But there's maybe certain decisions that maybe that guys would be better at**..." (woman, rock lobster fishery, in her forties).

The pervasiveness of resistance against women in industry decision making is similarly demonstrated when mandatory board quotas are considered. In this case, women are not derogatorily re/constructed on the basis of the behaviours and practices they adopt, but simply because they are *women* in a masculine domain.

> ".... people that are on boards ... think the **woman's there as a token**. 'We've set aside this position for a woman. She is the female representing the female community'...but the problem is everyone else then just automatically thinks she is the token woman and we can just **discount she's there for any credible reason"** (woman in fishing, key informant, in industry decision making).

Strategies

Women in fishing suggest many strategies to increase and facilitate women's participation in industry decision making including: financial support for travel, training and conference attendance; training and workshops on industry issues and decision making processes; formal Recognition of Prior Learning for skills of women in family fishing businesses; women being invited to participate; educating men; women's networks; mentors, role models, media profiling of women and awards for women; recognition of women's stakeholder status; and valuing *"difference"*, particularly by recognising women's knowledge.

"…[women] need to be shown that they have got information of value to impart, that they have got the ability to be a public figure in the industry. To be able to make a difference and to be able to share the knowledge that they have because **each person has a unique knowledge**, each person has a unique understanding" (woman, rock lobster fishery, in her forties).

Recognition of women as stakeholders

The need to recognise women in fishing as industry stakeholders was identified by several women.

"I think the industry has to accept them [women] and then the confidence will come….I think **men have to accept them"** (woman, rock lobster fishery, not involved in industry decision making, in her thirties).

Laying responsibility for cultural change on fisheries management agencies and industry organisations rather than with women in fishing, several women suggest that men in fishing require education on gender issues.

"It's **not just educating the women. It's educating the men** … that the **women do have a right** to have a hand in and have a role and their perspective might be different to theirs but it doesn't mean to say it is not of value" (woman, rock lobster fishery, in industry decision making, in her forties).

Laying this responsibility for cultural change with fisheries management agencies and industry organisations is also reflected in a focus group exchange where women feel they could be acknowledged as industry stakeholders and encouraged to participate by being invited.

Participant 1: "…women won't be involved until the government embraces them …because otherwise it's not going to happen"

Participant 2: "And they don't do that"

Participant 3: "No, they don't do it. It never happens"

Participant 4: "They **don't send an invitation for the women** and they never do. **They're sending it out to the skipper**"

Participant 2: "**You never ever get anything to say 'Well you are important to our industry'"** (women, rock lobster fishery, focus group)

National Rural Women's Awards and media profiling are also seen as positive strategies to foster recognition of women as industry stakeholders. However, some women in fishing are concerned that awards do not increase recognition for the "*average wife*" who contributes substantially but does not claim recognition or status-that is, they collude with a secondary industry positioning and do not have a "*big mouth*".

> "You can put a greater focus through the media on females….but **most of those were people like me…do a little bit more than the average female** does, and that's really sad **because the average wife… does a hell of a lot but …I don't know how you would … acknowledge them**…. You just get born with a big mouth and you're fine" (laughs) (woman, rock lobster fishery, in industry decision making, in her forties).

Industry organisations

Constitutional change in industry organisations to entitle women in fishing to full membership is also a practical strategy to increase recognition of women's stakeholder status and participation. Suggestions were also made that boards could address their selection processes and practices by freely advertising vacancies and having an independent person involved in setting criteria and selection processes to reduce the masculine privileging inherent in the "old boys" network.

While the majority of fishing industry boards do not have mandatory quotas or nominated positions for women, women expressed concern that mandatory quotas and nominated positions encourages women on boards to be perceived as "*token*" and lacking credibility. One woman in fishing felt it was more useful to have a less rigid target. However, she also recognises that locating women with the skills and motivation to be involved requires significant effort from a board. In a context where legislation does not require boards to increase their representation, it is unlikely that this effort will be made, particularly as women on boards threaten masculine power.

Education and training

Education and training for women in fishing on industry issues and decision making processes is a strategy to increase women's participation. The following extract asserts that increasing women's knowledge/ power through education and training is directly related to women's increased credibility. However, this implies that women currently lack knowledge (power). An alternate discursive reading and interpretation suggests that *legitimating* women's (*pre existing*) knowledge/ power through education and training may increase women's credibility.

> "If we can actively involve women [in training]... they ... almost will have to be listened to ... because but as people know, **knowledge is power** and if they've got the knowledge on how things work and are involved**... industry will need to stand up and listen to what they have to say**..." (male representative, fisheries management agency).

However, this representative goes on to explain that even the less confronting assertion (his assertion) that women know nothing, but may learn something through education and training and may gain knowledge/ power, is acknowledged as being an uneasy path for women and one which is likely to provoke resistance from men. Failing to recognise this resistance as a significant and/ or structural barrier, which he ironically indicates in some cases is so subtle that it is not spoken, once again places the final responsibility for involvement with women.

> "...there are a **number of dinosaurs** in this industry...and...men feel intimidated by women ... so **I expect there would be some resistance** but I think that from the people who ... have influence over the process, **if any of them do feel that way, they wouldn't be silly enough to actually say it** ... so whatever resistance there is, I don't think will be anything other than a small speed hump. But ...what **we've got to make sure is that the women in the industry themselves stand up** and take the opportunity because, **it's all good for men to say we'd love to have more involved,** and we want to run these training courses and etc etc, and **if the women themselves don't join in and participate...then it makes it a bit of a struggle"** (male representative, fisheries management agency).

There is also recognition by some women in fishing that education and training needs to be both ongoing and relevant to women's level of confidence and self recognition of their skills and knowledge.

> "... **you've got to put the carrot out so they want to come** ... and if initially you ... give an **education session on the bookwork** ... they'll come to that because 'there is something that's my area' and then **from there you ... build it up**... talk about the Lewin current and it's influence on the fishery" (women, rock lobster fishery, focus group).

Role models and mentors

Many women in fishing feel that role models and mentors are valuable. Some women also identify the value in role models and mentors as providing women with support and advice on strategies to counter and resist masculine exclusionary practices.

> "...you have to gain their [men's] respect and quite often it takes twice as long as any male, and then you could have a male sitting across the table ... who doesn't even know what is on the agenda for the day, who hasn't read anything, but ...he's the hero of the day. It gets back to that boys club bit, but ... **if those women...can come out of the meeting, go home, mull it over in their head, pick up the phone and ring someone...to sound off...like a mentor"** (woman in fishing, key informant, in industry decision making).

The pervasiveness of this masculine domination in industry decision making is evident in a comment by a male industry representative who, while espousing the virtues of role models and acknowledging the industry as *"male dominated"*, locates the responsibility on increasing the participation of women *with women,* without recognising the need for cultural change that would shift the responsibility for cultural change onto the fishing industry, currently recognised as the collective of (fisher)men, to ensure women are accepted.

> "... because it is a male dominated area at this point in time, there probably **needs to be ...pioneering...from a number of women**...who do pioneer the role that women can play in the industry and fisheries management decision process so that others can say 'Well they are doing it, why can't I? and ... start **to shift the [women's] thinking away from 'Oh it's all men...and that's just the way it is'"** (male representative, fisheries management agency).

Confidence and critical mass

Women feel that the strategies outlined above would increase women's confidence, recognition as industry stakeholders and representation in industry decision making. Some women also express the belief that women's participation will gradually expand in line with increasingly liberal broader social discourses of women's roles. Increasing the critical mass of women in industry organisations is also seen in itself as a mechanism which would encourage other women to participate and challenge the dominant discourse that industry decision making is the arena for (fisher)men.

> "It might just **jolt a few [men] into thinking 'Hey! I'm not just the fellow who has the oyster lease!'**, that **there is someone there as well"** (woman, oyster fishery, not involved in industry decision making, in her thirties).

However, the majority of women in fishing recognise the pervasiveness of masculine industry domination and feel that women will require significant and ongoing encouragement and support to increase their confidence and participation. Mechanisms to achieve this include collectives of women and exposure to subjugated knowledge.

Women's networks

Many women in fishing believe women's networks could increase women's confidence, representation at meetings, provide training and support, reduce isolation and provide an opportunity for the sharing of issues, problem solving and learning from other women. While supportive of the concept of a women's network, several women explain that their involvement would be restricted by similar barriers to their participation in mainstream organisations including the costs of training, conferences and organisational membership, travelling, child care issues, work and business commitments and a lack of time.

However, some women feel that women in fishing should increase their participation in mainstream organisations and that women are risking further alienation and ostracism through women's collectives. Several women also make the distinction between industry issues, where they feel that mainstream organisations are the appropriate forum, and support groups for women that foster women's confidence and knowledge of industry issues.

Collectives of women and alternate discourses

Women in fishing's collective action has the potential to increase women's representation in industry decision making, by advocating for women's access to, and the provision of, practical opportunities such as training and, by motivating women to political action through exposure to subjugated knowledges about gender. One woman describes an experience when she was involved with a course conducted through Australian Women in Agriculture (AWiA).

> "One [woman]...had level 10 school and had only ever had an entry level job.... We said to her 'What's your role in your family business?'
> and she said, 'I do the books...'
> and I said 'Are you a co-director in the business with your husband?'
> She said 'Oh no, **I just help in the family business'**.
> Then I said, 'Are you a company director?'
> You could see her face changing and then she said 'Yes. I am a company director'....I met up with her a few months later and **she said that that particular discussion that we had there had changed her view on her role in the business...It was a matter of how she saw herself, a perception**" (woman, oyster fishery, in industry decision making, in her fifties).

The impact of collective exposure to alternate discourses can impact on women's subjectivities and has the potential to motivate women to political action. This was demonstrated in this focus group exchange when women discussed a proposal by fisheries management to replace the family "camps" on the Abrolhos Islands with dormitories for "fishermen" for environmental reasons. This would mean that women and families would no longer be able to live on the islands with their partners during the fishing season.

> Participant 1: "I'm no better than anyone else. That issue was brought up about those dormitories and that should have been targeted more by the fishermen"
>
> Participant 2: "We are all at fault there"
>
> Participant 3: "We didn't even speak up!"
>
> Participant 1: "We think we are all very practical people and we just showed how stupid [we are]"
>
> Participant 3: "You know what we need? We need someone like you to get everyone together and then you can lead...but we need someone who can get people together"
>
> Participant 4: "If we had could have had something like this [focus group discussion] over the years... none of us talk to anybody else like we have talked this morning. I mean it's so different to be able to... and everybody knows what you are talking about.... I think it's been a great idea"

Participant 2: I think that we should get out and be seen more. **One thing that has come out of this [focus group] is that ... women are in the background without actually putting their hand up and standing up. They... let their partners do all the talking, decision making"** (women, rock lobster fishery, focus group)

One example of the power of women in fishing's collective action occurred when a group of women from the rock lobster fishery formed to lobby government in response to the "home porting ruling". This ruling required fisher(men) to return to the same port each day despite fishing from various areas along the coast. The consequence of this ruling meant that fisher(men) would spend more time away from their families. This women's group initiated action, including organising approximately 250 women and children to march on Parliament House, arranging social functions and lobbying government. Women in fishing's activism resulted in this ruling being overturned.

However, this women's group did not attend meetings, was only offered associate membership on the state peak industry organisation and dissolved soon after the issue was resolved. Women who had been involved in this group identified the masculine exclusionary practices that prevented this women's group from participating equitably in a broader range of industry issues.

Participant 1: "Associate members, **we weren't allowed to be full members** [of the state body]"

Participant 2: "'**Cause we were women**. No other reason"

Participant 3: "It's still very male dominated"

Participant 4: "We had to have just about marchers in the streets to join ... didn't we?"

Participant 1: "And they wrote back and said we could be associate members. **Like we could come along as long as our eyes and our mouths were closed. It would be fine it you didn't say anything.** So women, we've joined, but **we've never really got involved ... because ...their philosophy was wrong"** (women, rock lobster fishery, focus group)

Furthermore, this women's group was fighting for an outcome which also advantaged men and the industry in general. Women's activism was based on preserving the industry and family economic units in its current (masculine) form and was located within women's own positioning as secondary to "fishermen"-"***backing** what the guys are doing*". Thus, this activism perpetuates masculine power industry power relations.

> "It sort of showed **you're backing what the guys are doing**....you were **backing them in the end** basically" (woman, rock lobster fishery, in industry decision making, in her fifties).

Women's Industry Seafood Network Community

The Women's Industry Network Seafood Community (WINSC) is the most visible alternate discursive site for women. This organisation was established by women in fishing at a state level in 1996 and at a national level in 1999 to influence industry decision making for the betterment of the industry. The *WINSC Corporate Objectives and Strategic Priorities 2000-20003* (WINSC n.d.) describes the two original objectives of the organisation:

> We wanted to provide...an industry driven and industry managed women's network **so that the voices of the women of the industry could be heard.**
>
> We also wanted to provide a vehicle which would **promote the importance of women and their skills** and **raise awareness of the challenges faced by the seafood industry....**It needed to be owned and operated by its members, and they needed to play an integral **role in promoting the need for the seafood industry to discover its best untapped resource WOMEN.**

These statements indicate the organisation's intent to influence industry decision making, ensure women are heard, increase women's recognition and promote the fishing industry. However, these ideals may be competing. Conflict between the knowledge/ power of fisheries management agencies and fisher(men) dominates the industry. As such, in order to expose the alternate discourse of women as industry stakeholders, discursive space must be made, or taken from, within the existing dominant masculine industry discourse. It can be suggested that successful promotion and activism for the industry as a whole relies on unity within the dominant industry discourse and, in this context, it may be difficult for women to simultaneously support their industry and increase recognition for women as industry stakeholders. The construction of women in fishing as an untapped resource also implies that women are able to contribute more substantially to the industry by *increasing* their workload.

Since its conception WINSC has displayed considerable activism. This organisation has developed: the *Information Booklet in Meeting Roles and Procedures for Regional WIN* (Gill 1999), the *Charter of Best Practice: Seafood Women in Decision Making* (Smith and Sullivan 2001), a mentoring program; a social impact framework for restructuring fisheries; encouraged and commissioned research on women in the fishing industry; run workshops to break down barriers between fisheries management agencies and fishers; taken women and families to Parliament House to increase recognition of their role in the industry; provided training opportunities for women and changed a state industry organisation's constitution to have one WINSC member on the board, thus providing the opportunity for this representative to be nominated to the state peak industry organisation. One woman described how grass roots industry activism from this state WINSC group resulted in a 95% 'Yes' vote being returned by members in support of this constitutional change. WINSC has also had had representation on the Regional Women's Advisory Council (RWAC) and on the Rural Women's Secretariat (RWS).

WINSC addresses many barriers to women's participation in mainstream industry organisations. Diversity is enhanced by having a constitution where membership is open to anyone interested in the seafood industry including men, consumers, deckhands, and representatives from fisheries management agencies and industry organisations. WINSC also creates opportunities for participating that do not require physical presence at meetings. This targets women with small children and women with jobs outside their business.

Women involved in WINSC express a powerful alternate discourse advocating for the increased visibility and recognition of women as industry stakeholders.

> "...WIN either at a state or a national level has that role to play to ensure ... that when **government think about policy, they're not only thinking about the industry, they also are aware that there are women involved, and families involved, and communities involved"** (WINSC member, in industry decision making).

WINSC members speak about the importance of providing women with opportunities to network, mentors and role models, and educating and empowering women to encourage their recognition as industry stakeholders and increase their participation in industry decision making.

> "...**representation on boards, management, committees and industry associations** is the way to go. So try and **promote women into believing they are capable** ... **empowering** them...giving them the **skills and confidence** to do it....and **changing the culture** of the industry ... to change the focus from just an industry made up of men **to an industry made of women and men"** (WINSC member, in industry decision making).

However, WINSC has also experienced resistance. The perceived threat to masculine power is evident in the jocular suggestion by a (fisher)man (*"bloke")* that women will move to appropriate the practice of harvesting and relegate men to the less privileged home sphere. However, the amount of positive activism undertaken by this South Australian WINSC group targeted at industry issues (in contrast to gender issues) which does not breach masculine discursive practices, means that WINSC has received support.

> "...the first time I spoke to a bloke about it ... he said 'Oh well, **now I suppose you girls will want to go out in boats and we'll stay at home!**' ....but now he's one of our biggest supporters ..." (WINSC member, in decision making).

Conclusion

As in agriculture (Alston 2000), men dominate decision making processes and structures in the Australian commercial fishing industry. Few women are involved in industry leadership positions and attend meetings in their own right. Often when women do attend local industry organisation meetings, they accompany their partners but do not speak, or arrive to participate in social functions after meetings are concluded. Many women have also previously attended meetings and, as a result of this experience, have decided no longer to participate. It also cannot be discounted that some women who do not openly participate in industry decision making forums make a contribution to and may influence decision making processes in a less visible, secondary manner, such as being "secretary" to the Secretary, and women from European fishing families who attend meetings to assist their partners communicating.

However, there are significant barriers and exclusionary discursive practices to women's participation in industry decision making. The masculine privileging in industry decision making is perpetuated through the construction of fishing as a male occupation, fisher(men) as industry stakeholders and women as "fishermen's wives". Women's lack of (fishermen's) "knowledge" (power) means women lack the credibility to participate in industry decision making. This is a particularly pervasive barrier for women due to their exclusion from the practice of harvesting fish and from the opportunity to access this requisite knowledge, particularly given the embodied quality of such fishermen's knowledge.

Masculine privileging in decision making is evident across both the aquaculture and wild catch sectors in the Australian commercial fishing industry and is evident in the three case studies. Yet, this does not mean that some women in fishing are not interested in increasing their participation in industry decision making, and indeed, when women are faced with a threat to their family economic unit, they will demonstrate greater resistance and become involved despite the barriers. This is also evident for women in European fishing families, who may experience the greatest barriers due to the greater rigidness of cultural "wifing" and "mothering" discourses. The greater involvement of women in the oyster fishery case study research in board level industry decision making may be attributed to their higher level of education compared to women in the other two case studies.

While women in fishing do not participate in industry decision making or have an "invisible" involvement or assist in a secondary capacity, they do not threaten to appropriate the masculine privileging of industry decision making. Where women actively participate in decision making they experience resistance and are derogatorily re/constructed for breaching the dominant masculine industry discourse by both the majority of men and women in fishing. Women in industry decision making are often re/constructed in a derogatory manner and as emulating the masculine qualities and behaviours displayed by many men in industry decision making. However, in many cases women are forced to adopt these masculine practices in order to enter and remain in this arena and demonstrate to those in power, typically men, that they have the capacity to participate. Many women in fishing explained how they had wanted to or had attempted to become involved in industry decision making but had decided to withdraw, or chose not to become involved, as they felt unable to or did not wish to compete in this masculine culture.

Several women involved in decision making felt they had become empowered through this process. Furthermore, despite the barriers, many women in fishing who were not involved expressed a desire to participate in industry decision making and suggested strategies to increase women's representation. The alternate discourse generated by WINSC has the potential to influence women's subjectivities, empower women and enhance their capacity for resistance, and indeed, may be the viewed as a standpoint for women in fishing. However, WINSC also perpetuates the dominant masculine industry discourse to some degree in its alignment with the knowledge/ power of fishermen in providing a unified industry position in the ongoing conflict between fisheries management agencies and industry that has historically and currently characterises many fisheries. Thus, the collective action of women in fishing may also act, in some ways, to perpetuate the advantage of those in industry power-(fisher)men.

In the following chapter, I explore the contributions women in fishing make in their business, families and communities.

CHAPTER SIX

WOMEN'S CONTRIBUTIONS

Introduction

The previous two chapters provide evidence that women lack visibility and stakeholder status in the Australian commercial fishing industry, including in industry decision making. In order to establish a claim to stakeholder status for women in fishing in Australia, this chapter explores the contributions these women make to their businesses, families and communities.

Despite a lack of recognition, previous research has demonstrated that women make substantial contributions to their businesses, families and communities in agriculture (Alston 1995b) and in fisheries overseas (Lambeth et al 2002). The small amount of research conducted with women in fishing in Australia also suggests this is the case (Alston, Webb and Fisher 2000; Stella 1998).

Having established that the dominant masculine industry discourse confers occupational and industry stakeholder status to fisher(men) and relegates women to the secondary position of "fisherman's wife", I commence this chapter with an analysis of how women's discursive activities derive from meanings associated with the subject position of "fisherman's wife" and how some women resist this positioning. This is followed by a comprehensive overview of the contributions of women in the Australian commercial fishing industry in the key areas of their businesses, families and communities, which includes discussion on women's collusion with, and resistance to, the dominant masculine industry discourse. I conclude this chapter with a discussion on how women's contributions, or daily activities, are extensive enough to justify the recognition of women in fishing as industry stakeholders. Furthermore, women's daily activities also mean women may make a claim to partial and situated knowledge (Haraway 1988), which can contribute to a standpoint for women in fishing (Harding 1987). Indeed, it is the intent of this book to expose this subjugated knowledge (Foucault 1980).

Women's Contributions and the Dominant Discourse

The "job of wife and mother"

The dominant masculine industry discourse privileges fisher(men) with occupational and industry stakeholder status and relegates women to the secondary position of "fisherman's wife". Indeed, the majority of women gain entry into the fishing industry through marriage. In marrying a "fisherman", many women adopt/ are relegated to the subject position of "fisherman's wife". This subject position is discursively associated with certain activities. The normalising power of discourse is such that one woman who describes these activities as the *"job of wife and mother"* also alludes to the *mandatory nature* of these activities.

> "... **my job** is **first and foremost to be a wife and mother** to these guys ...that's something I really enjoy, but that's **something I've got to do"** (woman, rock lobster fishery, in her twenties, three young children).

Acceptable discursive activities for women ("fishermen's wives") are not recognised as vital fishing industry contributions but instead as an extension of "wifing" and "mothering" and what is "naturally" embodied in gendered constructions of "woman". Thus, the activities of women in fishing are not acknowledged as "work" and many women are not recognised as being part of the fishing industry. Reflecting the pervasiveness of the dominant masculine industry discourse, many women in fishing do not recognise their own contributions as significant and, in some cases, despite contributing their labour, do not even identify themselves as part of their family fishing businesses. Women who collude by engaging in "natural" acceptable discursive activities without resistance or demanding recognition are rewarded by being constructed as *"good" "wives" and "women"*.

> "...the majority of guys couldn't do what they do without their wives. A lot won't admit it but...behind the man is a **good woman"** (woman, rock lobster fishery, in her forties).

The *"job of wife and mother"* involves doing *"whatever needs to be done"* to support partners, families and businesses. The secondary positioning of women as wives and mothers is demonstrated in the comment of woman in the rock lobster fishery, who describes herself as "*just the 100% wife...the behind the scenes worker*". Indeed, "*whatever needs to be done*" may be seen to equate with anything partners did not do, could not do or would prefer not to do.

> **"We try and do as much as we possibly can. Obviously the boat's the men's thing**... Greg will ring up and say 'Can you pick up parts and put it on the carrier boat'...**that's what we're here for**.... my part is basically keeping the kids happy at home, making life easier for them by taking pressure off, getting bills paid and their washing done, their groceries sent over, **whatever they may need .... that's what we do"** (woman, rock lobster fishery, non legal partner).

Women in fishing "juggle" different aspects of life including working for the business, paid employment, family responsibilities and volunteer work. Demands on women in fishing are often great.

> "... I'd be sitting up til **one o clock in the morning** doing all the **paperwork** .... I had a **full time job,** I was on **three committees**, had **two teenage kids** and doing all the stuff that I'm doing for the **family business** ..." (woman, Cardea, in her fifties).

Doing "whatever needs to be done" and "juggling" these multiple tasks makes it difficult to separate women's activities into discrete spheres. This interconnection of tasks, combined with a lack of defined (occupational) status other than "wife", means that many women in fishing experience and describe their industry involvement as a "*lifestyle*".

> "It **never stops**. It's always **integrated across the social** and **every other aspect** of your life" (woman, oyster fishery, in her forties).

This integration of the contributions of women in fishing into different spheres means that it is difficult to accurately quantify women's labour. This is compounded by the seasonal nature of fishing. Furthermore, women who collude with the dominant masculine industry discourse may not identify their contributions as "work". So, while women involved in this research report spending between two and 60 hours a week working for their businesses, this may not be an accurate reflection.

While some women in fishing enjoy this *lifestyle,* other women are tired and express resentment due to the constancy and amount of work. As a result of their involvement, many women are multiskilled. In this context, "doing more than one thing at a time" is seen as an embodied natural quality of women and one which fisher(men) and fathers did not share.

The discursive division of labour

The discursive division of labour functions as a regulatory mechanism that normalises and makes natural the allocation of tasks associated with specific subject positions. Visible, public, manual, sea based, practical tasks necessary to harvest fish are normal and natural for "fishermen", while invisible, private, non fishing, land based tasks are normal and natural for "fishermen's wives".

> "…there are **typical roles** that are carried out and the **man** always goes and **catches the fish** and the **woman always stays at home** and looks after …the **bookkeeping**….traditionally it's the men that attend the functions … and the **woman stays at home and does the other duties**…so **quite often you don't see that they are actually there"** (woman in fishing, key informant, in industry decision making, wild catch).

This discursive division of labour reinforces the appropriate place for women in fishing and operates to ensure that the occupational privileging of (fisher)men is not appropriated.

> "They're not used to women being out there. It's almost **threatening**… to **see a woman out there doing the hard physical work. Women should be in the kitchen cooking dinner"** (woman, oyster fishery, in partnership, in her fifties).

Due to stricter cultural discourses of "wifing" and "mothering", the discursive division of labour confines women in European fishing families more rigidly to the physical place of home.

> "…he used to **get annoyed** and still does to this day **if he comes home and I'm not home**…after him being away …or even if it was a day. He **liked me** to be **home for him"** (woman, Cardea, in her fifties).

However, the *discursive* nature of the division of labour becomes evident when deconstruction exposes that women's activities are more diverse, with many women in fishing engaging in visible, public and/ or sea based "men's work" including decision making, manual work and harvesting.

> "In our fishery, they [women] catch and process, and … do all the admin stuff...I do **whatever has to be done**.... If that means that I'm catching or cleaning nets or mending nets, I can do that today, but I can go and speak to the Minister this afternoon…" (woman, key informant, in industry decision making, wild catch).

In other cases, women become redundant when family fishing businesses expand and "women's work" is appropriated by men.

> "I did anything that was needed…but when he got his own business **it was done by all the blokes"** (woman, rock lobster fishery, in decision making, European, in her fifties).

This fluidity within the discursive division of labour enables social, historical and economic changes to be incorporated at any one point in time, in order for the discursive division of labour to function as a flexible mechanism to preserve power for (fisher)men, irrespective of context.

Resistances

Some women in fishing enjoy and are content with engaging in the discursive activities associated with the subject position of "fisherman's wife". However, some women engaged in these activities reluctantly.

> "…he comes home and says **'I've worked so hard today'** and **I think 'yeah**, I've done all the bookwork. I've been sitting up there for 6 hours and doing taxation, wages and everything else, running around after kids and **you've had such a hard day?' Isn't that typical? A typical male attitude**….After 25 years of marriage, **you get used to it. It's just another job that a wife does"** (woman, oyster fishery, two children, in her forties).

After time, several women in fishing resisted by publicly articulating their own contributions as valuable. One woman described how she no longer referred to herself as a housewife, but instead as a company director. Additionally, some women actively resist their secondary positioning and lack of recognition with their partners.

> "…I was feeling that…I was being **taken for granted**….I take him aside when the people have gone away and say**, "It's not your business, it's our business and I would appreciate if my effort was recognised**. I'd like you to modify what you say"…He doesn't like it, of course.... I just

quietly say, **'And it's helped to have a working wife**'" (woman, oyster fishery, in partnership, in her fifties).

Breaches of discursive practice by women in fishing are actively monitored and acted on by other women, and particularly older women who are mothers of adult sons who fished. These women feel strongly that their sons should have a partner who dedicated themselves to supporting their sons and their family fishing business. The trade off, which daughters-in-law are supposed to be "*grateful for",* is that they do not have to "work" (gain paid employment). The discourse is so pervasive that even some younger women who recognise changes in broader gender discourses collude with these views.

"...maybe as time goes by **things will change** .... **daughter-in-laws would be saying to their daughters**... **you're not going to have to have the life that I had...to be completely available and open to some guy**...we want you to be able to go and do your own thing....but **whether or not it would be a good thing**... **I don't know .... I can see the mother-in-law's opinion** ....I have got the option of going 'I don't have to go back to work if I don't want to'.... so **I can understand mother-in-law's ...**going 'make sure you're available for your husband...**make sure you're grateful.... your husband working his arse off has allowed you to be able to do that'"** (woman, rock lobster fishery, three young children, in her twenties).

There is recognition by some women in fishing that changes in broader gender discourses may result in increasing opportunities for women to resist being subsumed into the unrecognised, secondary subject position of "fisherman's wife".

"...things are changing in the women's movement and ...they **[fishermen] realise** that they have actually got a pretty **good deal** if their **wives are willing to basically revolve their lives around their husband. Which is what they do"** (woman, rock lobster fishery, three young children, in her twenties).

Another woman felt that the broader women's movement has empowered younger women and that younger women are more likely to breach discursive practice by demanding recognition. She felt these women are more likely to risk the possible cost of a relationship separation.

"...women don't realise just how much they do put in and the men don't recognise it either, but that's changing ... because **younger women are demanding the recognition** and a lot **braver** .... **If they don't get it**, they just say, 'Blow it, **I'm out of here**....I'll go somewhere where ... what I do is recognised'" (woman, oyster fishery, in partnership, in her fifties).

Working in the Business

Harvesting fish and manual work

While the dominant masculine industry discourse privileges fishing as a masculine occupation, many women had and do engage in a diverse range of manual activities, including harvesting, processing and gear, boat and engine maintenance (Figure 6-1). While some women in fishing do not enjoy manual work, particularly those who already feel they are "*overloaded,*" many women find the experience of engaging in manual work challenging and empowering. Many women also make other significant practical contributions, including doing laundry for crews, organising boat stores, cooking, "gophering" and ordering parts.

Figure 6-1: Oyster farming woman checking racks, male employee assisting

Women's involvement is contingent on the fishery, geographic location, ethnicity, lifestage and women's subjectivities. In Cardea, reflecting the more rigid discourses of "wifing" and "mothering" and the prevalence of extended family fishing businesses, no women had been involved in manual work for their businesses. In the Western Australian Rock Lobster fishery, some women had worked as deckhands with their partners. In the Tasmanian oyster fishery all women had had some manual involvement on their oyster farms. However, despite this involvement, the majority of women in oyster farming indicated they were unaware of other women who worked manually. This can be attributed to the geographical dispersion of oyster farms and the discursive division of labour that constructs women's manual contributions as "helping" and invisible.

Women in fishing are also excluded from sites of masculine power-boats and wharves in the wild catch sector, geographical locations with reputations as "rough" places, and certain tasks. Even where women are able to access masculine sites of power, for many women the discursive division of labour assigns them to particular gendered "wifely", "womanly" activities such as cleaning when assisting with boat maintenance. This gendered task allocation is also evident in the experiences of some women who harvest fish. One woman describes how she engages in *"girl things"* as compensation for not being able to do all of the tasks that (fisher)men do, despite these tasks not being part of her job and being paid less than male crew members.

> "...you just make up for it in other ways, like you still do the **girl things** that **girls can do on a boat** like...**cook** ... **I didn't have to cook, it wasn't my job"** (woman, rock lobster fishery, fishes in family business, in her thirties).

Collusion by women and men in fishing is justified by natural embodied sex differences. However, the presupposition that (fisher)men have more physical strength and capacity than women in fishing cannot be attributed to an essentialist understanding that men embody more physical strength than women, as women's involvement in the industry is contingent on men *already being* in the industry. It can be assumed that fisher(men) have the physical strength required to fish and have trained their bodies to do so, whereas the physical strength of women in fishing will be more variable, as physical strength is not a prerequisite for being a "fisherman's wife". However, some women in fishing also have the capacity to develop the necessary strength required to harvest fish.

> "I was a **size 8 when I started** and **now I've got popeye arms. My body has just changed**....You **build up to it"** (woman, oyster fishery, working full time on the farm, in her twenties).

The masculine ownership of harvesting and physical tasks is also reflected in equipment design.

> "There are **no women size gloves or waders that fit a woman**. That is an industry that needs, is begging, to be opened. We need a fashion designer that can design waders to fit a girl" (woman, oyster fishery, in her twenties).

Many women in fishing also recognise that current practices are suited to the bodies of men (which exclude women) but that women could do things differently to facilitate their greater involvement in harvesting fish.

Women's involvement is influenced by their previous experience in primary industry and exposure to alternate discourses of equality, particularly from fathers. Their involvement is also determined by the financial need to preserve the family economic unit, particularly in newly established or acquired businesses. However, as family fishing businesses become more lucrative, regardless of women's desires, women who harvest are often displaced by paid (male) crew and relegated to more "wifely" duties as defined in the discursive division of labour.

> "...we try to help as much as we can....most of it **now is take morning teas rather than help with pots**....because **we have a deckie now** who comes and helps with all the gear..." (woman, rock lobster fishery, previously worked as a deckhand for partner, three children, in her thirties).

Women's involvement in harvesting is also influenced by their life stage. In the rock lobster fishery, women who had harvested fish with their partners tended to cease fishing when their children were born. However, several women harvested fish when they were pregnant and continued to take babies and small children fishing with them. This was particularly the case when there was a perceived need for women to continue harvesting fish.

> "...Jim was going to the Reef every day **on his own** and I just thought **if he goes out there and doesn't come home, I'm never going to know**... we are coming with you ....It's **scary stuff** ... your on like a little dinghy ... and you've got a beautiful angelic 18 month old baby in a bean bag ... and you're facing up to seven foot waves **... your heart is leaping out of your chest"** (woman, rock lobster fishery, in partnership, leasing license, in her fifties).

Women in fishing's participation in and enjoyment of manual work is influenced by the degree of adoption of broader social "wifing" and "mothering" discourses. Some women experience personal conflict as a result of the discursive tension between being a "good" mothers and a worker. Women in fishing who work manually from necessity often feel guilty about not caring for their children as they would like to, particularly in busy seasonal times. Reconciling this tension is also difficult for women who are planning families and want to continue working, particularly in rural areas where there is no childcare.

> "…I **still want to** be able to come to **work**…There's **no childcare**….**For me to stay involved the child would have to be involved.** Unfortunately Josh …went to an oyster farm … and heard funny noises coming out of a box…and there was a little baby in there. He thinks that's fine but **I don't know whether that is very good ethically to have a child sitting in a cardboard box in an oyster shed"** (woman, oyster fishery, no children, in her twenties).

Resistances

Many women in fishing feel that it is becoming increasingly acceptable for women to participate in manual labour due to broader shifting gender discourses about "traditional roles for women". One woman actively resists the discursive division of labour.

> "…**I'm usually the one that drives the boat**, because **that's what I choose** to do…. He'd like to but **he's not allowed to"** (woman, oyster fishery, second relationship, in her fifties).

However, many women face significant resistance from (fisher)men, particularly when they are "the boss".

> "…last Autumn I had a crew of four fellows with me.…they found it **really hard to work with a female** let alone…as **the boss"** (woman, oyster fishery, in partnership, in her fifties).

Women's involvement in harvesting fish also results in many women having significant knowledge. In some instances this is greater than their partners' knowledge.

> "I probably have **more working knowledge of what's happening in the farm. He quite often asks me** the questions of what's where, or what do

> we do now or whether we'll be able to" (woman, oyster fishery, in partnership, in her fifties).

Book keeping

Women's involvement in bookkeeping is extensive. The majority of women in fishing are solely responsible for all facets of bookwork and only in a minority of cases women share these responsibilities with their partners. In the few instances where women have no involvement, they have either recently married and are planning on taking on this role or are involved in extended family fishing businesses. Women do Business Activity Statements (BAS), taxation, liased with fisheries management and other government departments, pays for crew, insurance, quota transfers, loans, banking, paying bills, marketing, paperwork and superannuation.

As bookwork is not identified as "work", there is limited recognition for this activity and only a minority of women in fishing receive financial remuneration for this task.

> "[Industry women are] probably not recognised....Paperwork is not seen as work. Working in an office is not seen as work" (woman, oyster fishery, in her forties).

Where women do receive payment, this payment does not reflect the hours of work women invest and tends to reflect optimal taxation arrangements for businesses.

The discursive division of labour renders bookwork secondary to harvesting and, unlike harvesting, with the capacity for task bookwork not embodied in (fisher)men.

> "It's really invaluable, but they tend to be **behind the scenes**....a lot of women ...do the books, pay the bills, bank the money, do the BAS statements and income tax returns. Basically they're the accountant **behind the fishermen. Many fishermen are not attuned to bookwork, they're right into catching fish. Bookwork leaves them cold"** (male representative, oyster fishery, industry organisation).

Instead, bookkeeping primarily is seen as part of the job of being a "fisherman's wife" and women typically become responsible for this task after marriage. Women do the bookwork because their partners don't like it, or don't know how to do it. While this also may be partially attributable to the higher levels of education women have compared to their partners, (fisher)men remain privileged to choose their level of involvement.

One younger woman from Cardea describes how she is about to commence bookkeeping for the family fishing business. Reflecting the discursive supporting activities of "wife", bookkeeping is another task that she is expected to and prepared to take on *in addition* to her other tasks.

> "… I **don't know how much support I'll get as far as needing time out** to not cook dinner or not look after Davey **so that I can have the time to do that"** (woman, Cardea, one baby, in her twenties).

Women's involvement in bookwork is also contingent on the *need* for businesses to engage women in this task. Women are more likely to have a greater involvement in smaller businesses and in businesses that lack financial resources. When businesses expand, some women are displaced by paid employees irrespective of their level of enjoyment of bookwork and desire to continue. One oyster farming woman describes this experience of becoming "*redundant*" as "*traumatic*".

The power of the dominant masculine industry discourse to relegate/ encourage women to take on tasks associated with the subject position of "fishermen's wife", irrespective of women's capacity or desire, and to displace women irrespective of their desire, reflects and perpetuates the masculine occupational privileging of men and the primacy of harvesting fish and manual work.

One further theme to emerge is the substantial increase in administrative, regulatory and bookwork requirements, both in terms of the complexity of the tasks and the time bookwork requires, with some women in fishing estimating they have tripled the amount of time spent on these activities in recent years. Some women feel this provides an *opportunity* for women to become increasingly involved in their businesses, because their husbands lack the time, skills or desire. In some cases, increasing requirements means that women take over the bookwork which was previously done by their partners. One woman in the rock lobster fishery also recognises this as an opportunity for women to resist masculine industry domination.

> "I **nickname the fishing industry** actually **the Arab industry**….a lot of **females haven't been allowed to have their say**, which is unfortunate but they are **coming out of it now… the GST has bought women out**

because they have had to do the books where men have not wanted to learn … now it's got a lot more work, I'd say the women are getting into the bookwork" (woman, rock lobster fishery, in family trust, in her sixties).

Resistances

Many women in fishing collude with the dominant masculine industry discourse and engage in bookkeeping while receiving little or no recognition or remuneration. Many women justify this by employing the construct of "fairness".

"I couldn't go out in the boat and do what they do, but my husband also couldn't do what I do. So it's a **fair division of labour"** (women, rock lobster fishery, in her fifties).

Women in fishing do the bookwork for a variety of reasons: to save money; relieve stress on their partners; spend more time with their partners when they are home; to gain satisfaction from contributing; as they often have a higher level of education than their husbands; and in some cases, to retain some control over the family business. Women who do bookwork as a strategy to *"know what's going on"* and to maintain some degree of control, may be seen as resisting the dominant masculine industry discourse.

As women's contributions in the area of bookkeeping are not recognised as "work" but instead are constructed as a "natural" extension of "wifing", women's self recognition may also be seen as discursive resistance. Many women acknowledge their contributions as being extremely valuable and feel their businesses could not operate without them.

"I said to Michael, '**What would you do if I weren't here**?' He said, **'Sell'. If I were Michael… I would probably do that too**" (woman, oyster fishery, in partnership, in her forties).

One consequence of women's significant involvement in this area is their extensive knowledge of business operations. Many women report their own knowledge as substantially more than their partners, which in many cases is described as poor. Ironically, through collusion, some women in fishing gain power.

"Bruce took his passbook that I've got no name on to get some money out, because I just sign his name (laughs) and … said **'You've got to come in with me because they won't accept my signature'**….I had Sally from the National [Bank] ring me and she said 'Pat, **I've got a man standing in front of us** … described Bruce to a tee…and he's presented a cheque and **he says he's your husband.** And I said 'Well, he probably is'. Laughed my head off" (laughs)…. (women, rock lobster fishery, in partnership, in her fifties).

While women in fishing often take over the bookwork soon after marriage, there is resistance by mothers-in-law to relinquish this role, and these mothers-in-law often retain significant control over the extended family economic unit. This is particularly the case in European fishing families, where women are concerned about the potential dissolution of extended family businesses through divorce. This results in the exclusion of daughters-in-law from the family business and can be attributed to the desire of these women to preserve their economic family units.

"I know of quite a few women …they don't know what's owing, what's owed … and even some of their husbands wouldn't have any idea because their …**parents might not want to let go of the purse strings … and let go of the control"** (woman, rock lobster fishery, two children, in family trust, in her twenties).

Family business decision making

Women in fishing are marginalised in decision making in their family fishing businesses. While the few women who are sole owners have full autonomy, only a minority of women in partnerships have significant influence. It is common for women to make small day to day financial decisions. However, in the majority of cases, reflecting the occupational and industry privileging of (fisher)men, decisions relating to practical fishing matters such as the purchase of new boats or equipment tend to be made by (fisher)men.

"I think that it is that **historical thing. 'I'm the one that goes out on the boat and catches the fish. I'm the one that makes the decisions'"** (male representative, rock lobster industry organisation).

(Fisher)men are privileged to make decisions by virtue of their occupational and industry stakeholder status. The lack of recognition of women's contributions as "non working" "wife" results in the reduced

capacity for women to participate in decision making and justifies their exclusion.

> "...she **probably doesn't have much say** because **she's not going to work**..." (woman who fished, key informant, in industry decision making).

However, the pervasiveness of masculine domination in decision making in family fishing businesses is evident even where women receive recognition for their contributions.

> "... quite a few people...**talk about** the fact that **their wife works in the business** and that **she's the brains behind the business....I'm not saying that that means that she gets to make the decisions**..." (woman, wild catch, key informant, in industry decision making).

Some women in fishing collude with the dominant masculine industry discourse by positioning themselves as secondary.

> "We do discuss...how we'll spend the money....Although I have had input, **generally that's his area**. He has **more understanding** than I do **.... I'm just a labourer"** (woman, oyster fishery, in partnership, in her sixties).

Many women in fishing also collude by maintaining this positioning and explaining that they *"do what they are told",* even when they are not happy with decisions.

> "I...**let Bryan do that and be supportive of any decisions** he makes ... most probably over the years **I would have made some different decisions**.... he **never asks my permission**. He just says '**Sign this blank document (laugh) for the next loan'. 'OK'"** (woman, rock lobster fishery, ethnic, in her sixties).

This collusion may be attributed to women's reluctance to threaten their personal relationship and family economic unit by breaching the dominant masculine industry discourse.

> "... family businesses **wouldn't survive** unless there was **support** .... I **haven't tried to run the business** either. **If you have two people who don't agree then it could fall apart"** (woman, oyster fishery, in partnership, in her forties).

Women's contributions to and influence in decision making in their family fishing businesses is determined by several factors, including their subjectivities, partners, ethnicity and the involvement of extended family. Women tend to take on more responsibility for decision making when businesses are not established or are recently acquired and, in some cases, during peak fishing times. For some women, their bookkeeping contributions are part of a progression towards greater involvement in business decision making including exporting and marketing. Daughters-in-law in extended family businesses are excluded by older women, who maintain significant control of bookwork and related aspects of business management.

There was also diversity between case studies. In the oyster fishery, women are more involved in business decision making. However, in Cardea, apart from day to day regulatory and financial affairs, no women who participated in this research had significant input into business decision making. Ethnicity and age are exclusionary barriers for these women.

> "A lot of our fishermen are of **Mediterranean** extraction and **I doubt whether the women in those families would have much of a say**. But that's just a **culture thing** … Particularly as a lot of the **older generation** are still there....where there are **many Anglo-Saxon fishermen… women do play a bigger role in decision making** in the fishing enterprises…" (woman, key informant, wild catch, in industry decision making).

An additional barrier for these women is extended family involvement. One woman explained that it would not be practical for her family business, which involved her husband and his four brothers, to consult all of their wives.

Resistances

Resistances against women involved in decision making in family fishing businesses by men employed in related service industries such as fish sellers and bank managers are common. One woman in the oyster fishery, who is a sole business owner, described how she is ostracised by secondary service industry employees who construct her as aggressive, fearsome and masculine.

> "I'm very straight up and **people tend to be a little bit cautious** .... Someone said to me yesterday, **'Well everyone is just too scared to say anything to you'"** (woman, oyster fishery, owner operator, in her forties).

There are also several examples of women in family fishing business partnerships being ignored and not acknowledged by bank managers, accountants and marketing employees.

> "...even talking to a bank accountant...**because you are going in there without your husband** .... **women aren't recognised** ... they are always **expecting the better half**" (woman, rock lobster fishery, in partnership, in her thirties).

Some women in fishing demonstrate resistance in the face of this ostracism by withdrawing their business.

> "When I order stuff... **they'll say 'When can I ring back and talk to Ray?'** I'll say **'You're talking to me. You're not talking to him. I'm the one who's asking** for the stuff'.... **Some people deal with it** .... The **one's that don't, well I don't deal with any of them** because if they can't handle that, well then–tough" (woman, oyster fishery, in partnership, in her fifties).

It is also common for daughters-in-law to be excluded from business decision making. The following story reveals the emotional tension and conflicts between daughters-in-law, parents-in-law and partners in an extended family economic unit.

> "It's a partnership between the three of them. Alison and Mark [in laws], and Josh [husband]....it's been their life for 20 years but Josh and I definitely want to get our hands right into it....we've pretty much got autonomy in all decision-making at the moment but...on the financial side, some things...we don't know about....Gently, gently would be the word....**Emotionally, it's their daughter-in-law stepping in and taking control and, 'What is this woman doing to our family?'**....
>
> I do all our BAS and accounts, our daily expenditure and income and I find it very frustrating when it comes to end of year kind of tax....I don't know a lot of things....You've got to ask the right question ...and sometimes I don't get what I want....Josh and I would like to... say, 'Look, **we really want to take this over...but it hasn't happened**....
>
> I clash a bit with Mark, Josh's father...He's a bit old and he's 'She doesn't know what I know'....

...they volunteer all of their time and they don't get paid for any of it. Whenever we do need money, it's their money that tops us up....which also takes away our control in saying 'No, you're not working today', because the next day we might have so much work on, we have to have them. That's the tension....

It's a lot better now....and I can now say, 'Mark, that's not right'.... I figured it was age, because I was 21 when I started here....**they were probably thinking, 'Oh well, she's probably a bit young and isn't going to hang around for long'**....So there's all those kind of things as well ...**whether or not I was a good worker, 'Is she going to be a good daughter-in-law?'**....

To be honest Josh is the decision maker. **If a decision is made**, even though I've been the one who's pushed it, **I don't want to implement it....because of Alison and Mark**. I don't think they'd give it much credence or support if was just me alone" (woman oyster farmer, working on farm, tertiary educated, in her twenties).

In European fishing families, women are also ostracised by extended family members for possessing knowledge (power) and attempting to participate in business decision making. This is the experience of one woman when she attempts to become involved when her brothers-in-law come to their home to discuss the family fishing business.

"They've come here ... to do business, particularly quota transfer or purchase of gear ... and it's **fine while I'm making them a cup of coffee** and ...then Salvo says 'Sue, come here and do this', so **then I've got to start doing paperwork and that's not quite kosher**, because **I know about the business** and because I **understand more than even they do and I start to tell them.** And OK, and they're not quite sure what to say... **cause I'm not supposed to know about that stuff cause I'm a woman"** (woman, Cardea, in her fifties).

Some women in fishing experience resistance from their partners which function to keep women in their secondary position. However, several women resist their partners by employing strategies such as hiding money for the off season and *"being smarter"*.

"...if it's something that I really want to have done, **I'll just ring the accountant and the accountant will ring him and suggest it and then it's fine**, it'll happen... that's **just being tricky**, which housewives are.... quite often the information I give him, he won't believe until someone else gives it to him" (woman, rock lobster fishery, in industry decision making, in partnership, in her forties).

Three women had become involved in business decision making when they saw their family fishing businesses beginning to fail. Under these circumstances, it is acceptable for these women to assist, even as "the boss". However, recognising that being positioned in this powerful subject position threatens masculine power and potentially their family economic unit, despite enjoying this role, one woman in the oyster fishery described how she had ceased being the "boss" once her business became profitable.

> "…**he's the boss and I feel like I shouldn't take over**. So I told him I would do it **only temporarily**…. because it's my business and the business was collapsing and I had to do something about it…the business was going alright again and… I said to Fred 'It's **no good to have two bosses** in there', **because I work in one way and he works in another way.** So…**I decided not to go any more"** (woman, oyster fishery, in partnership, in her forties).

Several woman involved in WINSC participated in this research, and these women are promoting an alternate discourse about the contributions of women in fishing, their lack of recognition as industry stakeholders and women's right to be involved in business decision making.

> "Women have not been recognised. They've been the **invisible partners**…it is **always the man who has the final say….I'm trying to encourage women that it's a partnership agreement where they both have a say**…" (woman, key informant, WINSC member).

Paid Work Outside the Business

There is significant diversity amongst women in fishing in terms of their participation in paid employment outside of their businesses. The financial status of businesses, the age of children, rurality, education level, partners, ethnicity and women's subjectivities are all factors which influence the likelihood that women work outside their businesses. At the time of this research, only five women in the rock lobster fishery worked outside their businesses and 25 women received no other income than from their fishing businesses. Nine women from a total of 17 in the oyster fishery worked outside their businesses and in Cardea, two of a total of six women worked outside their businesses.

Women's paid work outside their family fishing businesses is highly dependent on the financial status of the family business. While older women tend not to have significant debt, many younger women did. These women are most likely to be working to assist businesses financially in servicing debts for purchasing fishing businesses, capital, licenses and leases and generating income for household bills and expenses. This is particularly the case in Tasmania, where oyster farms tend to be purchased rather than inherited. It is also necessary to some women to work due to income fluctuations which are contingent on regulatory changes, markets, environmental factors and stock fluctuations.

The majority of women in fishing who work outside their businesses also report working for personal reasons, including enjoyment, satisfaction, social contact, to keep up their skills, female companionship and to generate their own income and have their own independent interest. However, women who are providing essential income to struggling businesses are also likely to be providing significant amounts of unpaid labour in their family fishing businesses while juggling family responsibilities. These women are tired and stressed and some women explained that they would rather not have to work outside the home and business.

Some women in fishing enjoy not having to undertake paid work. In viewing their primary role as supporting their families and businesses, these women are most likely to collude with broader discourses of "wifing" and "mothering". However, some women, who would like to work, are also restricted by these discourses.

> "...a lot of [oyster] **farming wives get frustrated** and they **want to do something themselves to earn money.** By then you're **saddled with four kids**.... then you look on the fact that **'Well, this is my lot in life and this is my career to be a mum, to make a happy home'.** I think **as long as you accept that and try to convince yourself that it's a good thing, it works. Yes, you struggle at times though"** (woman, oyster fishery, in her sixties).

More rigid discourses of "wifing" and "mothering" in European fishing families result in some women not being *"allowed"* to work after they marry, despite wanting to work and the potential availability of childcare through extended family networks. Two of these women explained that if their partners died they would seek work for personal enjoyment.

Generally, industry discourses concerning women working outside their businesses are less regimented than relating to other areas. This can be attributed to it being equally acceptable and desirable for "fishermen's wives" to not work and be available for the business and family, as it is for "fishermen's wives" to work to assist financially when required. Despite greater discursive flexibility around women in fishing working outside their business, in several ways women's work remains secondary to that of men's. Many women forgo their careers and employment to marry into the fishing industry. Women with professional careers move into rural areas where there are few or no employment opportunities and childcare provisions to be with their partners. Women (and not men) cease paid employment for family reasons and to make time to provide unpaid labour for their businesses. One woman also describes how despite having tertiary qualifications and wanting to work she did not, as it would have adversely affected taxation of the family fishing business.

Women's paid employment is often required to fit in with family responsibilities. As such, it is likely to be part time or casual with lesser conditions than full time employment. Women also tend to work in traditionally feminised occupations including teaching, nursing and in service and hospitality sectors, areas which historically are poorly remunerated. Several women also work casually in the seafood processing sector with this providing one (scarce) employment opportunity for women in rural and regional communities. There is a gendered division of labour in this sector. Women work in roles which reflect their "natural" embodied capacity for manual dexterity, speed and their tolerance for repetitious work and standing still. In contrast, men in the processing sector work in management roles and in positions with greater variety and physical freedom. The predominance of repetitious work in the processing sector is reflected in one woman's estimate that the processing sector is approximately comprised of 80-90% women.

Women in fishing living in rural and regional areas, who work in order to support their businesses financially, are also likely to be employed in jobs with poor conditions which do not match their skill level. As well as a lack of suitable professional employment opportunities, women in rural and regional areas report additional difficulties with travel time and costs, and a lack of childcare. Accessing employment in rural areas is also difficult for older women without qualifications.

Resistances

Some women in fishing work outside their businesses despite significant resistance from their partners. This resistance is most likely to occur where women work for pleasure and when there is no financial need. In order for her to continue working, one woman in a European fishing family needs to ensure that she also fulfils her obligations as "wife" and "mother".

> **"My partner is against it. To take the time out from other things I should be doing** (laughs) **which is being a slave to him** (laughs)....You've got to really juggle a lot when you've working. Just so that you can't have anyone say 'Hey you're not doing that cause you're working'.... **it just puts more pressure on you.** You just work til 9 o'clock at night, make sure the wages are all done" (woman, rock lobster fishery, European, in her sixties).

Women in fishing who undertake paid employment are also sometimes ostracised by women in fishing who are not in paid employment. Again, while women work to support their businesses financially, this is viewed as necessary for their family economic unit and is acceptable. However, where women choose to work for personal reasons alone they are often ostracised and re/constructed as *"selfish"* and *"ungrateful"*, particularly by mothers-in-law. This was typical in European fishing families where there is a tradition of patrilineal inheritance and where mothers-in-law have an interest in ensuring their sons are adequately supported.

> "There's a **nice system that you get married and you put in really hard** and **they earn more money so you don't have to go out to work** (laughs) ....**if you're out at work doing your career** and your husband's doing his career, **you're not putting in**. Cause you can't do everything...[working is] **a selfish thing"** (woman, rock lobster fishery, mother-in-law, in her sixties).

Professional women who move into rural and regional areas to be with their partners and who continue working for personal reasons also experience significant resistance from women in fishing, who are often local women with lower education levels.

> "It's difficult because again the **wives don't understand why I need to work,** have a career. They **don't understand the ... they way I want to do it**... even though they might be our age, they've already got 12 year

olds. Just **different lives"** (woman, rock lobster fishery, tertiary educated, in her thirties).

In some cases, professional women are ostracised by these local women in fishing and re/constructed as "lesser" mothers.

"I would **never put my kids in day care**....**to go off to work**.... a lot of women that go working whinge all the time. So I think, **'Well what do you want?' .... I'd sooner do the job that I do good**..." (woman, rock lobster fishery, three teenage children, European, in her forties).

Working in the Home

Reflecting the discursive relegation/ adoption of tasks associated with the broader discourse of "wifing", the majority of women in fishing are solely responsible for housework. While many partners assist if asked, this is often done reluctantly. Only a few women have partners who contribute without being asked. While several women make household decisions jointly with partners, many women are solely responsible for household decision making and household finances. Fishing practices, including unpredictable schedules and the seasonal nature of fishing, mean that partners are often absent during busy household times and sometimes for days at a time.

Many women in fishing collude with broader discourses and the dominant masculine industry discourse that reflects and perpetuates women's contributions as "non work".

"I don't work, so as far as I'm concerned **that's my role"** (woman, rock lobster fishery, three teenage children, in her forties).

Women's collusion also perpetuates the discourse of "fairness" in the discursive division of labour, where women in fishing are at home while their partners harvest fish.

"I'm the **home body** and this is my home and **where I work ....so I don't expect him to do a lot** around the house" (woman, rock lobster fishery, in partnership, two children, tertiary educated, in her forties).

Being at home with children and being responsible for households is seen both as a privilege and an equal trade off for having a partner who harvests fish and provides financial support.

In European fishing families more rigid discourses of "wifing" and "mothering" mean there are more clearly defined expectations about the "job of wife". Several women in fishing attribute these expectations to European culture and the upbringing of their partners.

> "....**housework** and hanging out the washing and that sort of **stuff is not his job. And that's the way that he was raised"** (woman, Cardea, in her fifties).

Two women from Cardea with Anglo Saxon backgrounds who had married into European fishing families, describe how difficulties in meeting these expectations had led them to suffer depression. The following extract also reveals how women collectively perpetuate and reinforce the dominant discourse.

> "...his mother stayed home and cleaned. So **he expects a clean house and he expects I should be able to do all that** by myself.... that's **a fairly Italian sort of [thing]... I try. It's not in my make up.** Honestly ....I think it's frustrating that I don't do it [keep the house clean]. I mean, **us mothers** speak about this ... and we **say if we left the boys at home with the kids...you'd expect meals cooked and the house cleaned.** You would**... even though we know what it's really like**" (woman, Cardea, one child, in her twenties).

The power of mothers-in-law operates to ensure that daughters-in-law are meeting the discursive standard and providing adequate care for their sons. These mothers-in-law also transfer similar expectations for care to their sons.

> " (marrying into an Italian family) was a **whole new culture**.... it was a learning curve, but **I was very lucky** that ... he was the first one married, so.... he **didn't have a sister or sister-in-law that I had to measure up to, so I basically set the standard, and his Mum.... was wary of me** ...but she saw I could clean a house and look after kids and **she taught me to cook**....None of this baked lamb crap for them, they didn't even want to know what it was" (woman, Cardea, in her fifties).

Given this pressure, it is understandable that many daughters-in-law collude to avoid the consequences of breaching discursive practices. One woman from Cardea was socially ostracised and re/constructed as a "*slack woman*" and "*dreadful housewife*" by her partner and extended family for employing a cleaner, thus breaching discursive practice. However, in the face of this hostility she was able to resist by asserting her Anglo Saxon background.

> "I'm a **slack woman**. I'm a **dreadful housewife**. I used to get told I was the **worst Italian house wife in Cardea**, but then I remembered I'm not Italian (laughs). So I started saying to them **'Well I'm not Italian, what do you expect?'"** (woman, Cardea, in her fifties).

A further point of discursive deviation was that she had exposed her sons to an alternate discourse by encouraging and expecting them to be independent. The consequence for this breach was being re/constructed as a "*bad mother*".

> "I'm **not a very good Italian mother**. I don't dote on my sons....I love them dearly. But I'm not going to be around for ever and if they can't stand on their own two feet by the time they're about 16, tough....**That's another one of those things that makes me an enigma**..." (woman, Cardea, in her fifties).

The demands on women in fishing to provide the discursive level of care means many women enjoy the time when their partners are absent harvesting fish.

> "He's not at all domestic and it's **like having another child** living in the house. It's **quite pleasant not to have to look after someone** for those couple of days..." (woman, oyster fishery, in partnership, in her forties).

However, in some cases, resistances by partners operate to remind women in fishing of their "place" even when their partners are absent.

> "...he left the other day and [I said] **'Oh good, I don't have to clean the house...'** He just sort of laughed and ... [said] **'I need a clean house. If you need a break from it I guess well I'm not here well that's ok'"** (woman, Cardea, one child, in her twenties).

A lack of recognition of women's activities as "work" is also a form of resistance employed by partners. In the following extract, when the husband of a woman from Cardea is forced to be responsible for childcare and household tasks, he defines these activities as "*bullshit*". This construction reinforces her daily activities as of no value and clearly demarcates his role as non domestic in the discursive division of labour.

> "... there are times when he's had to take over .... And **he stayed at home and took over ...my role**...and he finds that...he's very busy.... Having

to take them to different things after school … and **he thinks it's bullshit. That what he calls it!"** (woman, Cardea, four children, in her forties).

While some women in fishing continue to ask their partners for assistance with household tasks in the face of resistance-"*huffing and puffing*"-many women are resigned to the fact their partners would not help and that it was *"not worth"* asking. One woman in the rock lobster fishery describes not only resistance from her partner in terms of her being responsible for all household tasks, but his further power in obstructing her in these tasks.

> **"I do [the housework] and the cooking and the cleaning and the washing and the ironing**….He [husband] comes in and says **'Why isn't that done?'**….I'd actually like him to get off my back basically…. **I just want to do it when I can fit it in as opposed to when he wants it done**….he's quite **obstructionist** ….**if I'm doing something, he'll want me to do something else that revolves around him"** (woman, rock lobster fishery, three children, in industry decision making, in her forties).

Some women in fishing collude strongly with the dominant discourse and ostracise women who did not.

> "I see **a lot of fisherman have to come home and cook, or clean**. To me **that's not on**…. I **wouldn't expect him** be out on the ocean such long hours and come home and help me in the house**…. That's my bit. …He's got to go to work**…" (woman, rock lobster fishery, three children, European, in her forties).

However, after time, several women demonstrated increasing resistance to discursive prescriptions and their adoption of these tasks. One woman describes how exposure to subjugated knowledge altered her subjectivity and facilitated a change in her activities.

> "I think it's **an age thing**…a **rite…of passage** where you no longer have the responsibility of children…. If you put 20 years in bringing up kids and compromise all the other things that you wanted to do, **it then becomes your time in life. Some people see it as being selfish, but it's not**….the men think that you've gone nuts…. they have been so used to three meals on the table and **you give yourself the permission to have the 'I don't want to cook tea tonight" …. they're not very accepting of it**, but **you need to say to them, 'I don't feel like cooking tea tonight, so I'm not going to'**…. this is **written about** …in all the books… it's **not just me"** (woman, oyster fishery, in her fifties).

Ironically, colluding with the discourse also means that some women in fishing are able to gain power in household decision making.

> "… Sam was **never here** when the kids were growing up **so I had to make that decision whether he liked it or not** because he just wasn't here. And as far **as the household is concerned…his thing is 'I just want to go fishing and everything else you take care of'"** (woman, Cardea, in her fifties).

Family responsibilities

Women in fishing are responsible for caring for children, grandchildren and other relatives, maintaining family contact by extensive travelling, social planning, mediating family conflict and providing emotional support. Women's adoption of these responsibilities are influenced by unpredictable fishing patterns and the absence of partners and derive from broader discourses of "wifing" and "mothering" and the associated "job of wife".

> "... **my job was to look after the family**. Not that it was ever said, this is your job, this is my job. It was just the way it worked out….I made it as easy as I could for him….so **I handled everything that had to be handled at home while he was away"** (woman, Cardea, in her fifties).

The construction of these activities as "natural" for wives/ mothers/ women means that some women receive little or no recognition for their contributions.

> "I **don't think that…the husband probably realises just what's involved in trying to keep a family together and look after kids** … as well as actually trying to run the business and I know one situation where **the husband will come in and just expect that … he can just come up and put his feet up because she's been doing nothing**… and he expects her to do all the running around and 'Will you take this load of fish?' …'Can you get all the stores ready?', and 'You do this' … and she's ringing her hair out thinking …'Look after the kids so I can have five minutes peace!'" (woman, key informant, wild catch, in industry decision making).

While many women in fishing collude with this discourse some women resist by ceasing certain activities.

> "Shaun would be **away for the kids birthdays so …[I'd] get a present** and wrap it up…**After a while I thought, this is ridiculous**, I am creating this relationship thing with the children…**he's got to take responsibility** for putting something into it, finding out when their birthday is….**The presents just stopped coming"** (woman, rock lobster fishery, two children, in her forties).

Demands on women in fishing to care for their family are great, particularly when their partners are absent. Women have to take full responsibility for caring for sick children. One woman spoke about the difficulty of caring for her daughter with a mental illness for three years with her partner away for most of this time. Several women explained that their partners were fishing when their children were born. However, these women felt that this was acceptable as their husbands were "working", with one woman further justifying her partner's absence from the birth by constructing him as a "professional". This reflects the dominant discourse where (fisher)men's privilege to harvest takes primacy over all else. Several women also describe the difficulties in being both *"mother and father"* to their children.

> "[I was] **Mum and Dad**….all the **sporting events** and even… **childbirth….You didn't have your husband there**….he **didn't even have a day off.** So it was like 'I'll just drop you at the door, (laughs) and I'll go to work. Have good luck having that baby'…. But…I'd always been brought up that that was the way it was …and **with cray fishing.... you've really got to give it 100%**....You've got to be the **disciplinarian**. It's no good when a child does something wrong on Monday and 'when I see your father on the weekend'….You have got to…do something about it….Yeah, **its extremely difficult"** (woman, rock lobster fishery, European, in her sixties).

Maintaining family contact

Women in the Tasmanian oyster fishery and women in the rock lobster fishery travelled extensively to be with their partners and to maintain family contact. In Tasmania, the majority of oyster farms are approximately two hours drive away from the family residence. Women and children travel to their farms on weekends and during school holidays to be with their partners.

In the rock lobster fishery, men often fish from different areas during the season and women often travel to coastal areas where their partners are fishing. Many women also move permanently to coastal villages or to the

Abrolhos Islands for the four months fishing occurs here during the season, to be with their partners. However, reflecting the impact of broader discourses of "mothering", the majority of these women move back to Perth when their children reach high school age to ensure their children have access to quality education and greater diversity in social and other activities. While this is seen as necessary, this family separation impacts negatively on relationships as well as on the broader community.

> "...one of the biggest problems is education....we don't have the educational support up here for the high school ....**A lot of women are moving down to Perth and leaving their husbands here**, which is definitely **not a healthy option** and that's definitely **having an impact on family structures and town structures** and its becoming men up here on their own for 8 months of the year and their partner might come up once a month so they can see their kids" (woman, rock lobster fishery, coastal town, tertiary educated, two children, in her thirties).

The Abrolhos Islands

The Abrolhos Islands are a chain of 122 small islands on the mid West Coast of Western Australia located 55 kilometres west of Geraldton and extending 100 kilometres up the coast line. When fishing commenced here early in the 20th century, (fisher)men left their families on the mainland and lived in primitive "camps" when fishing was permitted from the islands. Gradually women and children started coming out to live on the islands to be with their partners. Currently, fishing is permissible from the islands for approximately four months of the rock lobster fishing season and many families live together in family owned "camps" during these months (see Figure 6-2).

Living on the islands can only be described as unique. While living conditions have dramatically improved and families now have generator powered electricity and improved communication, transport and hygiene facilities, the environment remains isolated.

Residing in this isolated environment means there are significant demands on women. Women have to be vigilant about the safety of young children who are not allowed outside without wearing lifejackets. Women either choose to home school their children or send them, often by boat, to the Island school. Women do bookwork using electronic communication. Again, reflecting the prevalence of broader discourses of "mothering" where women *"put the children's education first",* when children reach high school age women tend to remain on the mainland.

Figure 6-2: *"Roma Island in the Easter Group Abrolhos Islands"* n.d., although I was informed this was taken in "recent" years. Geraldton Public Library Photograph Collection. Photograph number 1245

Many women enjoy the months they spend annually at their island "camps" when families are together in this unique natural and social environment. However, many women also feel that it is unpleasant and isolating, with several women describing negative impacts on their personal relationships. Some women also describe the community on the Islands as *"closed"*. "*Newcomers*" and women who do not enjoy having daily morning coffee (complying with existing discursive practices), are ostracised and re/constructed as "*different*" and "*outsiders*".

> "If you haven't lived here for 25 years, you're an **outsider,** and **a lot of the fishermen and their wives have been born and bred here** … they've done everything together, their families. **And that sort of environment is… quite forbidding … to outsiders**…" (woman, rock lobster fishery, second relationship, no legal status, in her sixties).

Women also discuss the tension between managing the competition between family fishing businesses in the fishery and the requirement to maintain social relations while living together so remotely.

> "…out there it's **dog eat dog**. Everybody is **competing for the same lobster** ….You get a lot of the **blokes competitiveness** coming out … and unfortunately that …[can be] **transferred to the partners and families**…. so there is a degree of fragmentation …but **everybody loosely is a**

community and they work together … because out there you are miles from the mainland and **if something happens, you need somebody"** (woman, rock lobster fishery, in her fifties).

Social responsibilities

Women in fishing also have the responsibility of managing the family's social life. In Cardea, as fishing is dependent on weather, women are unsure of when their partners will be home and describe the difficulties associated with planning and committing to social events. Many women in the rock lobster fishery describe having no social life during the eight month season as partners typically only have a few days off each season including Christmas Day, New Years Day, Easter, and Good Friday. Women describe how their partners often miss important social functions and, when they do attend, are often tired and irritable. When partners do attend, women often provide extensive support to enable this to occur.

> "… we must attend this wedding, it's a very special family member …. **so I would … have … nice hot water in a bucket, all his clothes, his toiletries, his towel.** I'll have it all hanging in the car, I'll **meet him at Deepwater**, he'd have his shower … get dressed and off **we'd go to Perth, go the wedding, turn around and come back to Deepwater,** get out of all his gear and put his fishing gear on, hop in and **he's out in the boat to go fishing again**…That's always up to the woman, **all the driving**…"(woman, rock lobster fishery, in her fifties)

Most women in fishing feel their lives revolve around their husband's work life, with many women sacrificing their own social life and experiencing isolation. Several women also will not, or are "not allowed to", attend social functions if their partner is not able to attend, particularly in ethnic fishing families.

> "... you got a party … you **don't know what time you're going to get there**. You **don't know if you're going to get there**…. We **have to say no** to a lot of things, and **I won't go without my husband"** (woman, rock lobster fishery, three children, European, in her forties)

Mediating family conflict

Women in fishing also mediate family conflict which occurs due to the tensions of family and economic units being one and the same.

"Love and money. The emotional and the economic side of it often don't mix" (woman, oyster fishery, in family trust, in her fifties)

This conflict typically occurs between adult sons who are working together, and between sons and parents, where adult sons want more control over the business which parents are reluctant to relinquish.

"There was an attempt at the beginning to have a family business but it just did not work. It was very painful. **The head of the family wanted to remain the head of the family and the boys of course wanted to do their own thing** and the only way to sort that was to dissolve it" (woman, rock lobster fishery, in her sixties)

This conflict often results in significant tension for women in their personal relationships.

"…you **can't sack your family** (laughs) and that sometimes is **extremely difficult when you've got teenagers on board** …. Then you've got the **mother saying 'don't you sack my son'** (laughs). **The poor father's right in the middle of the whole thing"** (woman, rock lobster fishery, European, in her sixties)

The requirement to adjust

A major theme that emerged relates to the difficulties women in fishing experience in adjusting when partners come home after being away, after seasons have finished and when partners retire. Women often *"drop everything"* to make themselves available to their partners when they are home.

"When the men were home **we dropped everything**…. and **your life then is just living for your husband, 'cause he's home**….we have to be like that…. if it was bad weather for a few days, he'd come home so …any plans we had made …. **everything got dropped because Dad was home** and so that's how we lived" (woman, Cardea, in her fifties)

Women feel that they have *"two separate lives"* and many women reported *"changing"* relationships.

"In off and on seasons it's a **completely different marriage**….I basically keep my mouth shut for most of the time. **I wouldn't say things to him that I normally would… and basically whatever he needed would come first**. Even if it was something small like getting dinner early so he could … go to bed and making sure the kids were in bed by 7 30 and being quiet.

> He gets very tired … I wouldn't have any chores around the house for him to do….I wouldn't even tell him about it …. **I can play the whole wifey role. When he's around for four months he's able to do things"** (woman, rock lobster fishery, three children, in her twenties)

This adjustment means many women in fishing have to reconcile alternate identities of secondary, supporting "wife" and empowered, independent woman, with many women also expressing resentment about having to change. One woman explains how women's experience of independence can result in relationship breakdowns.

> "You see quite a **lot of divorces** … because **women end up being so used to being by themselves that they just can't slip back into a relationship** and quite a lot of them **don't want to because they've done all this work for 8 months of the year and their husbands come home and tell them that they can't watch that TV program** and…they want a proper meal … and it doesn't work like that anymore. … **it's a big thing for women to be on their own …and for us to try and accept them back into the relationship**….and quite often the adjustment period lasts almost the whole amount of time that they're home" (laughs) (woman, rock lobster fishery, three children, in her forties)

Emotional support

Many women in fishing discuss the importance of emotionally supporting their partners. This is especially the case for partners who fish away seasonally where women take full responsibility for caring for the family and running the household. Women also support their partners by respecting the informal rule that catches and fishing locations are not to be revealed to other women in fishing, who are often friends. This is particularly the case in the rock lobster fishery, where there is no quota which determines overall catches for the season. Women also worry about their partner's safety and several women spoke of the high accident and death rate in the industry.

Worry and emotional support also relates to concern for their partners in times of financial instability, industry fluctuations and management changes. Several women in fishing report that both they and their partners are stressed by issues including: drought which affects fish stocks in inland fisheries; the SARS outbreak which reduced export orders; the war with Iraq which resulted in the cancellation of orders for countries not sympathetic with Australia's involvement; uncertainty over fishing rights; and significant stress resulting from increasing administrative demands which, if completed incorrectly, can result in prosecution. In the face of these pressures, families have to maintain business operating costs as well as family living.

Women's contributions in providing emotional support are significant and many women describe how taking full responsibility for "home" and everything else reduces stress on their partners, which allows them to fish more safely.

> "… [**women**] **play a huge part** really. See men can physically go out and do that but **if it wasn't for the support at home**, it would be **very difficult being a fisherman** and concentrate 100% on the task…. to actually have your mind on something else at home could be a dangerous situation for fishermen" (woman, rock lobster fishery, two children, in her forties)

Working in the Community

The majority of women in fishing involved in this research are extensively involved in voluntary work in school, preschool, sporting and community organisations. They assist with various activities including reading, canteen, fund raising, school excursions and as committee members. One woman described how voluntary work changes with varying life stages of women.

> "It started with **nursing mothers group** … **your roles change with your children… as they grow through different stages of their life, you change as well.** But I was involved in **preschool things…mothers club,** raising money for the **school**…in the last 15 years… we have raised a lot of **money for local charities** here in the town …" (woman, Cardea, in her fifties)

Women in fishing who live in rural areas and older women are likely to be involved in voluntary work that support their communities, such as environmental groups, hospital auxiliaries, church groups and fundraising and lobbying to increase the infrastructure capacity of their towns. Many women are involved in multiple voluntary activities and the most significant barrier to women's further involvement is a lack of time.

The acceptability and prevalence of women's involvement in voluntary work derives from broader gender discourses which construct women as "naturally" caring, patient, supporting and as helpers. This construction of "natural" gendered embodied qualities reflects the lesser value given to these (female) voluntary activities at a broader social level. Voluntary work is unpaid and, in this research, was not considered as "work". Many women in fishing, who make extensive contributions and who are involved in multiple activities, feel they do *"not do much"*.

Voluntary work is seen by women with tertiary education in rural and regional areas as an opportunity to contribute and develop their skills. The majority of women also report that voluntary work reduces their social isolation, with two women describing it as a "*lifesaver*". Indeed, in some cases, voluntary work as an acceptable discursive practice for women is a *legitimate* way for women to engage in social contact.

In some cases, voluntary work also facilitates women's empowerment. This is particularly the case in Cardea where voluntary work enables some women to engage in *public* activities without breaching stricter discourses that operate to keep women in the home.

> "... we've been alluding to **... my husband's relations here who stayed at home and cleaned house**.... we set up a Committee called Cardea Italian committee.... They raise money for charities and a lot **of these women who've basically been down trodden and told they can't do anything except cook meals and have babies** suddenly **found that they were actually very capable** ....And it's been a **real privilege watching these women blossom** and take on responsibilities" (woman, Cardea, in her fifties)

However, the discourse relegating women to "home" is so pervasive that some women are not able to participate in voluntary work as their partners require them to be home.

> "I **used to** do the tuckshop roster ... **but having your husband being away** from home so much... **I couldn't tie myself down** to anything, **because you could bet the day I would ... he'd come home"** (woman, Cardea, in her fifties)

Several women are also involved in voluntary work that supports the fishing industry. In addition to women contributing to industry associations discussed in the previous chapter, this includes involvement in voluntary work that reinforces fishing community identity. One woman is involved in an organisation which aims to improve public perception of the fishing industry in her small town, where the industry is threatened by people who had recently moved into the area. Women are also involved in Blessing of the Fleet festivals. In one fishing community, in 1948 there were two committees to organise the Blessing of the Fleet celebrations, an annual event held at the commencement of the fishing season where boats and fishermen are blessed. One was a "men's" committee, responsible for tasks such as choosing the boat to be Blessed, and a "ladies" committee (Figure 6-3) responsible for catering, decorations and dresses for the girls. The pervasiveness of masculine industry domination which ascribes and encourages women and men in fishing to adopt certain subject positions and activities is such that these separate committees and allocation of tasks remain to this day.

Figure 6-3: *"First Women's Committee, Blessing of the Fleet" 1948* Courtesy of Fremantle Library WA Local history collection Photograph number 2654

Women in fishing make substantial contributions through their voluntary work in sustaining families, businesses and communities. Building infrastructure in rural areas such as educational facilities means that families are able to stay together and that children can access services, and improving public industry perception of the sustains businesses and the broader industry.

Conclusion

As in agriculture (Alston 1995b), women in fishing contribute extensively in a myriad of ways to sustaining their families, businesses, communities and industry. They provide labour in their family fishing businesses including manual work, "gophering", buying stores, cooking, laundry, book work and business management. In addition, many women work outside their businesses to contribute financially to their households and businesses. Women are also primarily responsible for maintaining households and caring for families and are extensively involved in voluntary work.

The most significant points of difference amongst the case studies are that women in European fishing families are more likely to adhere more rigidly to stricter cultural discourses of "wifing" and "mothering" which tend to confine these women to their homes. When these women breach discursive practice, they are likely to harshly re/constructed and socially ostracised. This greater confining to the physical place of home also accounts for the fact that these women are not as likely to work as women in the oyster fishery, who tend to have a higher level of education, greater debt and be involved in businesses which are less likely to include extended family members.

Drawing from broader social discourses of "wifing" and "mothering", the dominant masculine industry discourse ties the activities women in fishing engage in to the subject position of "fisherman's wife". As a result of marrying a man who is remunerated for harvesting fish, many women adopt/ are relegated to the subject position of "fisherman's wife". They place their own aspirations secondary and often undertake (and are expected to undertake) *"whatever has to be done"* to support their families, businesses and communities.

These activities are normalised for women through regulatory mechanisms such as the discursive division of labour and explained through women's "natural" difference and embodied capacity for caring, nurturing and supporting. As a result, these activities are not acknowledged as "work" or recognised as part of "the industry" and women receive little, if any, recognition.

However, as this chapter has demonstrated, women's contributions are vital to sustaining businesses, the industry, families and communities. Doing *"whatever needs to be done"* is potentially an infinite workload and women "juggle", prioritise and multi-task a range of activities. As a result, women's activities are not easily separated into discrete spheres. Indeed, women themselves do not separate their activities into distinct spheres of business, industry, family and community but experience and describe their industry involvement as *"lifestyle"*. Recognising the theoretical premises of the importance of internal factors including emotions in women's empowerment (Deveaux 1994) and valuing women's own experiences (Harding 1987), many women enjoy these activities.

However, some women undertake these activities unwillingly, and women who breach discursive practice by asserting their right for recognition, or women who do not engage in "natural" practices are derogatorily re/constructed and socially ostracised.

I argue that as a result of many women engaging in their daily activities, women have a claim to industry stakeholder status. Furthermore, I also argue that as a result of engaging in their daily activities, women can claim significant knowledge. However, as these activities are tasks which women "naturally" do as part of the (secondary) "job of wife"/ mother/ woman, women in fishing have no basis for their knowledge (power) to be legitimised. This stands in contrast to the knowledge/ power of men in fishing, (fisher)men, which is legitimated through their occupational status and experience and the privileged embodied "knowing" this enables them to claim.

The dominant masculine industry discourse may be interrogated and deconstructed to reveal a series of dualistic constructions which demonstrate their hierarchical and relational nature (Beasley 1999): man/ woman; primary/ secondary; "fisherman"/ "fisherman's wife"; worker/ non worker; occupational status/ "job of wife"; stakeholders/ not stakeholders; and knowledgeable/ not knowledgeable. These discursive constructions perpetuate women's industry marginalisation, the privileging of (fisher)men with occupational and stakeholder status and masculine industry power relations.

In the following chapter, I examine the implications of women's secondary positioning within the context of a changing industry and the implications of women's knowledge to enhance the future of the fishing industry.

CHAPTER SEVEN

WOMEN IN AN INDUSTRY IN "CRISIS"

Introduction

In the previous chapters I have established that women in fishing lack visibility within a dominant masculine industry discourse, despite making significant contributions to their industry, businesses, families and communities. This chapter explores the implications of the current industry climate for women in fishing.

This chapter commences with a brief overview of the current status of the fisheries in each case study fishery. I then proceed to deconstruct the language prevalent in the current dominant management discourse, including the language of "sustainability", to explore the political agenda evident within the seemingly positive and apolitical management discourse and the effects of how this power is exercised (Foucault 1980; Phoca and Wright 1999). This is followed by an assessment of how well management applies its own definition of sustainability–ESD. Exposing the knowledge many women gain from their daily experiences by privileging the "view from below" (Harding 1987), I then explore women's understanding of "sustainability" and the social impacts of management decisions. I conclude with a discussion on the potential for women's subjugated knowledge/ power to contribute to the future of the industry, the potential for this knowledge to form the basis of a standpoint (Harding 2004a) for women in fishing, and the likelihood that this will occur.

Status of fisheries in the case study sites

This research occurred at a time when both farming (Bourke and Lockie 2001) and wild catch fishing (Kura et al 2004) were acknowledged as being in "crisis". The quest to achieve a more sustainable industry in Australia is resulting in increasingly stringent State and Commonwealth fisheries management regulation (Caton and McLoughlin, 2000). The increasing management which is necessary to implement and enforce increasing regulation and the subsequent cost recovery policies have resulted in the escalation of costs required to operate in fisheries. The pressure to increase economic efficiency, in conjunction with technological advancements in boats and fishing gear, has also resulted in increased catches and declining stock in many fisheries, including rock lobster, blue fin tuna and the trawl fishery. This further perpetuates the need for increased regulation. In this context, the structure of the industry is changing. Commercial fishing is becoming increasingly corporatised, the number of owner operator family businesses is decreasing and aquaculture is emerging as a major growth sector (FRDC 2000, 2001; NAC 2001).

At the time of this research, the current sustainability discourse underpinning natural resource management in Australia was Ecologically Sustainable Development (ESD) which espouses the integration of environmental, economic and social factors. Fisheries management is also underpinned by a bio-economic philosophy in the broader climate of a neo liberal state, and as such, employs the privatisation of the rights to resources, primarily through the allocation of quota, as a mechanism to achieve greater industry sustainability. This process reflects the centralisation of those rights according to capital ownership-an area where women are marginalised (Munk-Madsen 1998a).

A further factor influencing fisheries management in the context of a neo liberal state is Nation Competition Policy (NCP), a policy agreed to in 1995 by the Council of Australian Governments (COAG) which includes representation from Commonwealth and all State governments. NCP involves "...a comprehensive package of reforms based on the proposition that competition, if properly harnessed, can provide substantial benefits for consumers and boost economic performance" (NCP 2002, v).

The Western Australian Rock Lobster Fishery

At the time of this research, the Western Australian Rock Lobster Fishery was undergoing a management review required by NCP. Until recently, the fishery had been managed by input controls, including seasonal restrictions and a maximum (150) and minimum (63) number of pots per license. However, during the course of this research research, three aspects of fisheries management were being reviewed-the lifting of the minimum and maximum number of pots per license, and going to an output control management system through the allocation of quota. In 2003, just prior to fieldwork commencing, the 150 maximum pots per license rule was lifted. This decision was made by the State Minister for fisheries, and during 2007, the Minister was due to make a final decision about "going to quota". A social impact assessment of these proposed changes indicated that they would result in significant negative social impacts for families and communities, and that the majority of industry did not want these changes to occur (Huddleston 2003).

"Going to quota" involves a process of allocating an amount of quota per license. Allocation can occur by equating kilos for current pot holdings or alternatively on catch history. Under both scenarios, there was the potential for some businesses in certain circumstances to be marginalised. Allocation by equating kilos for pots means that fisher(men) with above average catches for the number of pots they own, and who need to maintain this level of catch to remain economically viable, would be disadvantaged as they will have to purchase quota from other fisher(men) to maintain their necessary catch level. Allocation by recent catch history would disadvantage fisher(men) who, for any number of reasons, for example, ill health, have been catching less than the average catch for the number of pots they own and less than they may catch in another season.

Trawling and longlining fisheries in Cardea

One of the major fisheries operating in Cardea was the Commonwealth Trawl Fishery (previously named the South East Fishery). This fishery was one of the first fisheries internationally to become managed by ITQs in the early 1990s. Since this time, the fishery has undergone significant change: family fishing businesses have been marginalised and fishing effort and catches have increased. During this research, women spoke about the increasingly unviable nature of their businesses, yet they felt unable to exit the industry as they could not reclaim their investment in fishing boats and capital, and in the current regulated climate, it is extremely difficult to sell boats, which are often worth around one millions dollars. In late 2006, AFMA announced a major Structural Adjustment Package designed to "encourage" fishing businesses to exit the industry for Commonwealth fisheries. Local newspaper articles reported anger from the fishing community over inadequate compensation packages.

Tasmanian oyster fishery

While there was not deemed to be a "crisis" in the Tasmanian oyster fishery at the time of this research, the fishery's management agency employs ESD to ensure the development of the fishery remains "sustainable". This was occurring by encouraging operators to adopt Environmental Management Systems (EMSs), quality systems which may be implemented to an entire business or any aspect of a business. Alarmingly, despite ESD being a fundamental fisheries management principle in the aquaculture sector, in the context of a fishery not in crisis-that is, requiring immediate action-awareness of social aspects of sustainability are low. However, in the broader neo liberal context, the privatisation of resources could have significant impacts for the oyster fishery, and several women in the fishery during fieldwork were concerned about the future impacts of privatising water resources for up-stream agricultural users, and the quality of water at river mouths where oyster leases are located.

Deconstructing the Discourse of Fisheries Management

Deconstructing the language and exploring associated meanings which are currently prevalent in dominant fisheries management discourse reveals the political nature of a seemingly apolitical agenda and the effects of how this power is being exercised (Foucault 1980; Phoca and Wright 1999).

The term "crisis" can be read as the language of fear, and implies the need for someone who "Knows"-in this case fisheries management agencies-to take control and implement urgent action. Deconstructing the dominant fisheries management discourse reveals the positioning of blame for the crisis with fisher(men) – the "crisis" is a result of overfishing by fisher(men). Furthermore, there is a need for fisher(men) to be managed (controlled) in order that they will change their practices. The inference is that fisher(men) are not competent to manage, or contribute to managing the crisis.

Prominent in fisheries management discourse are terms such as "crisis", "pressure", "competition", "professionalism", "cost recovery" and "economic efficiency". These terms are used in conjunction with the language of "sustainability"–Ecologically Sustainable Development, Environmental Management Systems and a multitude of other similar terms which refer to implementing ESD strategies in particular contexts. This emerging language is so complex that the Fisheries ESD website (ESDRA 2005) explains that a committee was formed to clarify the meanings of terms relating to sustainability. This complex language highlights not only the need to *explore and expose* possible re/interpretations of meaning, but also the processes through which this language and meaning is *conveyed* from fisheries management agencies to industry operators.

Exposing meaning

Deconstructing this language reveals a premise that, in order to "increase efficiency", fishers need to become increasingly "professional". Consequently, there is an emerging subject position of "professional fishermen". The language of "professionalism" can be read as historically associated with "white collar" (male dominated) educated occupations in contrast to "blue collar" work (male dominated) characterised by physical and often dirty, hard work. The gradual incorporation of "professional" into the subject position of "fishermen" infers a formalising of occupational status and implies "professional" behaviours, skills and knowledge. This is significant in two ways. Firstly, the implications for women are substantial. I have already established that women are excluded from the occupation of fishing. This transition from "fisherman" to "professional fisherman" further marginalises women, as they have an additional and more formal layer of occupational status to achieve, above and beyond that of "fisherman".

Secondly, the re/construction of "fishermen" as "professional" can be read as exclusionary, where fisher(men) who are not "professional" no longer belong in the industry. Further interrogation of the meaning of "professional fishermen" reveals an association between this emerging subject position and a requirement that these fishermen are economically efficient, educated, technologically adept and able to keep up with, understand and implement the rapidly changing, highly complex and increasing amount of management regulations. The meaning exposed in this discursive deconstruction is that "professional fishermen" are fisher(men) who are able to engage with, and respond to, management in a bureaucratic manner and, that those who lack these skills no longer have a place in the industry. Several women in this research expressed concern that some fisher(men), particularly those who are older and of ethnic descent, lack the education, literacy levels, skills and knowledge which is required for the subject position of "professional fisherman". Indeed, some women spoke of fisher(men) who had exited the industry for these reasons.

The tensions for fisher(men) inherent in this shifting meaning associated with the subject position of "fisherman" is apparent in the Eastern Tuna and Billfish Fishery: Industry Code of Practice for Responsible Fishing (Jusseit and Robinson 2003) which suggests that fishermen need to maintain a "professional" image and therefore should avoid being intoxicated on wharves and around harbours.

There is also meaning implied in the constant use of the terms "*more*" and "*increasingly*", which are often used in the context of the need for fishing businesses to become "*more*" and "*increasingly*" professional and economically efficient. The application of a critical approach to linguistic meaning, where meaning cannot be guaranteed but is interpreted, reflects the ambiguity in, and the power exercised through, these terms. In addition to the fact that the terms "professional" and "economically efficient" are used repeatedly without a specific explanation given to their meaning, how much "more" and "increasingly" "professional" and "economically efficient" do fisher(men) need to become? Presumably, there is some divergence of understanding between management and 2nd, 3rd and 4th generation fisher(men) on both of these issues. Furthermore, in the immediacy that the language of "crisis" implies, there is also pressure on fisher(men) to attain sufficient professionalism and economic efficiency *quickly*. In this context, certain fisher(men) are disadvantaged as they may have a different understanding of these terms and may not be able to immediately achieve enough professionalism or economic efficiency.

This deconstruction reveals the hidden political nature of a seemingly apolitical management system-an agenda to alter the industry structure by reducing the number of smaller family fishing businesses by fostering increased industry corporatisaton in the quest for more economically efficient (sustainable?) fisheries.

Conveying meaning

In addition to deconstruction which exposes how power is exercised through language and meaning and its effects, exploring the manner in which language and intended meaning is *conveyed* also exposes how power is exercised, and its effects. The majority of women in this research discussed the increasing amount of written text they receive which outlines regulation changes and the substantial penalties for not complying with these changes. Several women discussed instances where the increasing amount and complexity of text meant that some fisher(men) were not able to grasp the intended meaning and so unintentionally did not comply and were penalised. A further factor which disadvantages these fishing businesses is the speed of fisheries management processes, again which may be justified by the need for immediate action in the context of "crisis", whereby these changes are expected to be complied with *immediately*.

Women and men in family fishing businesses with no experience of negotiating or interpreting text in the way it is intended are deemed as *inadequate* and *not competent* to be operating in the fisheries by fisheries management agencies if they are not able to keep up with the highly complex and swiftly executed regulatory changes. The problem is seen as an individual problem of fisher(men) who "lack" and who no longer belong in the industry. In this attribution of responsibility, no attention is given to structural issues. Yet, I would argue that management is ethically obligated to ensure that their requirements are equally accessible to all fishing businesses.

It can be assumed that larger, more corporate fishing business structures have a greater capacity then smaller family fishing businesses to quickly interpret complex and continually changing regulatory controls framed in fisheries management jargon, perhaps through greater financial resources which enable them to employ specialised accountants, lawyers and scientists.

Therefore, in the current industry climate, it is increasingly important for fishing businesses to operate on par with bureaucratic fisheries management agencies practices. The extract below, which refers to events at a local industry organisation meeting, reveals that this task is challenging for one woman with tertiary qualifications and years of fishing industry board experience. This provides an insight into the greater disadvantage experienced by European fishing families and older fisher(men) and their partners. These management practices expose the agenda to displace smaller family fishing businesses.

> "... just at our last meeting **the fishermen** were sitting there with their head in their hands **saying that they've no idea what some of the arrangements are** because ... AFMA kept changing them.... and people were just completely lost... but **then even I found it hard to understand**. I threw my hands up in the air and said 'I don't know what applies to what species!'" (woman, wild catch, in industry decision making)

Consultation processes are also an avenue for conveying meaning between fishers and management and are discussed in the following section.

"Us and them"

Fundamental to understanding women's potential to enhance the future of the fishing industry is an exploration of the hostile relationship between "fishermen" and management (Roespstorff 2000). This was evident in the prolific references by women in this research to women and men in fishing as "us", and to fisheries management as "them". This relationship can be attributed to issues resulting from the privileging of scientific knowledge over the experiential knowledge of fisher(men) (Palsson 2000) and the increasing shift in power from industry to fisheries management agencies evident in the increasing legislation and regulation of the industry since the 1950s (NCP 2002). The majority of women in this research feel that fisheries management agencies make decisions based on a different and less accurate knowledge than fisher(men).

While fisheries management agencies operate on the premise that unregulated fishers will inevitably fish stocks to extinction, many women in family fishing businesses challenge this bio-economic assumption on the basis that maintaining sustainable stock levels is essential to maintaining their livelihood. Several women refuted the notion that lesser regulation would result in the exploitation of stocks to extinction.

> **"Rape and pillage** the earth? Well, I think **it's like a farmer shooting all his stock** each year. **We don't"** (woman, wild catch, in industry decision making)

Several participants pointed out that there are many areas where families have been fishing for generations without regulation, and that this could not have occurred if fisher(men) had not employed sustainable harvesting practices.

> "They don't go out to rape and pillage cause **if they did they wouldn't be able to fish next year....** That's **why they are still there as fourth and fifth generation fishers....** if their grandfathers had done that, they wouldn't be able to be doing it now" (woman, wild catch, in industry decision making)

Many women recognise the need for some level of industry management. One key informant woman described how "self regulating" fisher(men) in her area recognised the need to change their harvesting practices and voluntarily developed a management plan. Fisheries management had adopted this plan and there was no conflict in this fishery between industry operators and management. This supports previous research (Zwanenburg, King and Fanning 2000) in demonstrating the potential for successful management practices to emerge from an inclusive, "grass roots" process which incorporates the "local" knowledge of fishers. This is contrast to a management process dictated and implemented from a remote, centralised bureaucracy and based only on scientific "expert" knowledge.

In many fisheries, there is significant industry resistance to fisheries management agencies, who are often perceived as unethically dictating fishing practices and lifestyles.

> "... the state government said that they wanted to ensure that the fish was caught in the most ... cost effective manner but ... **since when has that got anything to do with government? That's industry's decision how they choose to catch the fish**.... the issue is ...non exploitation of resource but in terms of cost that's an individual's decision.... So **to have a power coming in and dictating who can and can't based on how cost effective they are is a real dilemma"** (woman, wild catch, in industry decision making)

The struggle for the privileging of knowledge between fishers and management was a theme raised by many women who participated in this research, most prevalently in the rock lobster fishery and Cardea, where fisheries were undergoing significant change. While women's feelings about management in the oyster fishery ranged from negativity to ambiguity, in the rock lobster fishery and in Cardea, many women exhibited a comprehensive knowledge of industry issues and were scathing of current fisheries management practice.

Consultation processes in these fisheries are deemed to be poor and there is an overwhelming feeling that fisheries management agencies make decisions without regard to the views of family fishing businesses. Generally, it was felt that fisheries management agencies consult because they are required to and, that final decisions are already made prior to this consultation occurring. This research provides some evidence that fisheries management consultation and communication processes are inadequate. Notice of pending consultation could be improved, as there are many instances where women report learning of regulatory changes through the media. Several women also feel that it is difficult to access information from fisheries management agencies. Consultation processes are also gendered. The rock lobster mailing list for consultation processes is comprised of "stakeholders". While this list is potentially open to all stakeholders, a fisheries management agency representative acknowledged that anyone other than a Managed Fishery License holder is required to "*come forward and identify themselves as interested*". This favours men who are privileged with industry ownership and excludes women with no ownership status.

There are also further factors which influence the capacity of women and men in fishing to participate in consultation processes. For example, in the season prior to fieldwork in the rock lobster fishery, the beach price for lobster had dropped from $36.00 a kilo to $10.00 a kilo. Many women and men in fishing had been working hard and there was considerable concern in the industry about the viability of many businesses continuing should there be a similar price fluctuation during the following season.

A further challenge for fisheries management agencies is their lack of capacity to cater for and incorporate diverse views in consultation processes. Rather than management recognising the value and benefits inherent in a diversity of viewpoints, and revising their approach to consultation to better incorporate diversity, the following quote implies fisheries management agencies take the stance that the diversity of industry views are irreconcilable or "too hard", and as such, they need to make an executive decision. Current fisheries management practice exacerbates the tension between fisheries management agencies and men and women in fishing.

> Participant 4: "That one was just that **we heard it and that's what was happening"**
>
> Participant 3:" It was **in the paper and that was it. Other than that we get no notification on that 150 pot rule"**
>
> Participant 1:"This lifting the 150 pot rule has been talked about....after a certain time, the fishermen just think 'Ok well that's been talked about but nothing is happening about it'...**suddenly it's a reality and ...they are doing exactly the same with this quota"**

> Participant 3: "In the **paper** last week… saying that it is well and truly on the cards that **we will have quota**"
> Participant 4: "But why should it happen? This is where **I get angry. Why should it happen if the majority don't want it? That's wrong and that's where our government steps in"**
>
> Participant 1: "I think **government does what they want to do regardless** and they keep splashing out on meetings … but **it always comes back to what government want**"
>
> Participant 4: "So … **we need to be a body and stand up** and say we do not want that but very seldom do you ever get all of the fishermen together to stand up and say we don't want that**, some will say they don't, some will say I don't know** ... and **that's when they bring it all home"** (women, rock lobster fishery, focus group)

A contrasting perspective on fisheries management consultation processes is provided in a fisheries management report by Fletcher et al (2005).

> "Consultation on management of the rock lobster fishery is conducted in a **open, accountable and inclusive** environment where **all sectors of the industry** and the Departments Managers **collectively identify and discuss appropriate courses of action**. Decision makers take **due note of advice provided** on the basis of this consultation and give **reasons for decisions which vary** from consultation-based advice….The consultation process **is extremely well understood** with relatively high levels of participation from the **various stakeholder** groups" (Fletcher et al 2005, 82)

However, yet another perspective was provided by fisheries management agency representatives, who in interviews, described industry as largely apathetic.

> "…generally...**industry are apathetic**....broadly speaking...they regard the management process as something that someone else takes care of …" (male representative, fisheries management agency, rock lobster fishery)

This construction of an "apathetic" industry is associated with industry's minimal engagement in consultation and decision making processes. The term "apathy" also implies a "lack of caring", which, when positioned in the context of "crisis", reinforces the notion that fisheries management agencies must take control and act. In contrast, my own conclusion after interviewing the women involved in this research, is that there is not one woman who "does not care" about her industry. For example, women in the rock lobster fishery were concerned about the impacts of the changes occurring under the NCP review, including the inevitability of some operators being disadvantaged in the allocation of quota. Several women in Cardea were concerned about the future of their industry, and indicated they would sell their boats and businesses if they could, but that industry insecurity meant that no one would buy these assets. Several women in the oyster fishery were concerned about water quality issues and the impacts for their oyster leases of allocating water resources for up-stream agricultural users.

Deconstructing "apathy" within the dominant fisheries management discourse suggests that fisheries management agencies consider women and men in fishing are not participating in the way which fisheries management deem fit and/ or, that fisheries management agencies are not interested in genuine consultation. An alternate reading of an "apathetic industry" suggests that, if fisheries management agency consultation processes are inadequate, women and men in fishing may feel they cannot "keep up" with fisheries management, or if consultation is conducted without the genuine intent of gaining feedback, women and men in fishing may feel there is simply no point.

> "I whinge ... but there's **not very much you can do** …[fishermen] **don't know how to go about it** …. If you went to the fisheries **you wouldn't know who to talk to**. You're **not ever asked to give your opinion** as such. If you go to a meeting...there's maybe 100 other people there that have a different view....they get a **range of view points but...they can't spend the whole evening discussing everybody's"** (woman, rock lobster fishery, three children, in partnership, in her thirties)

In the fishing industry, poor participation and perceived "apathy" may be reinterpreted as disempowerment. In consultation processes, once again, larger, corporate fishing businesses with financial resources and greater access to "expert" knowledge are advantaged, and indeed, are more likely to be heard and able to exercise more power in fisheries management agency decision making processes. One male fisheries management agency representative explained that management consultation and decision making would be easier and more cost effective with fewer operators. This exposes the agenda of fisheries management agencies to displace smaller, family fishing businesses in the broader context of a neo liberal state.

> "… the level of investment in the fishery and the shift from owner operator …is basically driving the fishery to be more efficient .... hard nosed and more professional … the **culture of owner operator**, which is basically someone fishing for their family and their family business...because of the life style, **will change to it being a corporatised fishery** where the ownership is concentrated into the hands of large fishing businesses … who are driving the fishery to be operated and worked and managed on a basis that provides the best economic return for their business ….**we can manage ecologically as well against that scenario as well with an owner operator scenario**. People could even argue that it is **easier to do** it [under a] commercialised, industrialised type of arrangement **because you have less people to deal with in the decision making processes and therefore you might be able to spend less"** (male representative, management, rock lobster fishery)

In the conflict between management ("them") and fishers ("us"), women are united with and support their partners. This reflects the power of management to threaten the survival of family economic units and the level of industry consolidation required to resist this threat. In the context of fisher(men)'s and management's struggle for privileged knowledge, women in fishing are less likely to seek, or be granted, discursive space for their views. There is no room for women to claim privileged knowledge relating to a new and competing discourse that challenges women's lack of stakeholder status. If women were to claim this knowledge in the current climate, they would be diluting the industry resistance to fisheries management, and entering into competition for discursive space with their partners. For some women, this could jeopardise the power they exercise in their family economic units. Additionally, the potential costs of the discursive breach of women making claims to knowledge and stakeholder status could pose a threat to the survival of their family economic units.

Women have demonstrated collective resistance and exposed their subjugated knowledge, for example, when women in the rock lobster fishery fought to overturn the home porting ruling discussed in Chapter 5 in the section Collectives of women and alternate discourses. However, this is not resistance to a masculine dominated fishing industry, but resistance to a fisheries management decision threatening their family economic units. Women's resistance reflects the strength of their feelings about maintaining their lifestyle and family economic units.

In the struggle for the privileging of knowledge between fishermen (us) and management (them), possibilities for women to both claim stakeholder status and to have a voice are further reduced. In this context, women will not be granted discursive space to expose an alternate discourse and any assertion by women for such a space puts them in potential conflict with their (fisher)men partners. Thus, women collude with and perpetuate the dominant masculine industry discourse by supporting their (fisher)men partners in resisting management when their family economic unit is threatened.

(Fisher)men can make a "local" knowledge claim which is legitimated through generations of practical fishing experience, and fisheries management agencies can make an "expert scientific" knowledge claim which is legitimated through educational qualifications and organisational status. However, as women's knowledge derives from the "natural" activities of being a woman, wife and mother, women are further disadvantaged as they have no basis on which to substantiate their claims to knowledge. Thus, exploring women's understanding of sustainability and their experiences of the social impacts of fisheries management decisions exposes women's subjugated knowledge. Prior to this, it is useful to explore how fisheries management agencies incorporate the "social" in sustainable.

Management and the "Social" in Sustainable

The prioritising of economic factors and the environmental imperative of ensuring renewable stocks over social issues was reflected in discussions with representatives from industry organisations and fisheries management agencies. These representatives typically either did not mention social issues or positioned the "social" in sustainable as secondary to environmental and economic concerns. For example, management and industry representatives, as well as a minority of women in fishing, suggested that social issues could only be considered *after* the industry was "sustainable". One fisheries management agency representative asserted that the rock lobster fishery was only able to consider "this stuff"-social issues-as a consequence of the fishery *already* being managed sustainably.

> "the fishery has been **so successful in managing its sustainability**, [it] has meant it has **the opportunity for this stuff [social issues] to be discussed**...it would be a lot harder to talk about social issues ...if we were facing the gun barrel of a lobster stock collapse" (male representative, management, rock lobster fishery)

The social impacts of fisheries management agency decisions have rarely been considered and many women in fishing expressed concern about this. Even when social impacts were specifically raised with fisheries management agency representatives by women and men in fishing, social issues were still accorded a secondary status.

> "... its hard work to get it across and if we challenge the managers to say 'You're **not taking into account the regional implications of this move'**. They'll say 'Yes we have but **the ecological objective is the paramount one'"** (woman, wild catch, in decision making)

Indeed a philosophy was revealed in some discussions that social and ecological issues are mutually exclusive and irreconcilably incompatible. One woman involved in decision making feels it is dangerous to legislate for *"a social thing"* for this reason.

> **"AFMA doesn't have a social objective in its legislation**....I wouldn't be arguing for a social objective ... because **once you put the social thing in there its very hard then to manage for the sustainability for stocks**. ... you get to be like Canada and you're paying billions of dollars for social welfare for out of work fishermen or you manage your stock into extinction to look after the families" (woman, wild catch, in industry decision making)

This separation between social, environmental and economic factors in "sustainability" fails to incorporate the intended/ espoused *integration* of these factors. Furthermore, this *disintegrated* approach to achieving a "more sustainable" industry reflects the Western, masculine, scientific reductionist discourse characterising fisheries management.

The awareness of fisheries management agency and industry organisation representatives of the importance of social issues varied between case studies. Where there is no gain to be made from considering social issues, fisheries management agency and industry organisation representatives display a lack of understanding of social issues within the context of ESD. For example, when responding to a question about how the oyster fishery incorporated social issues in sustainable fisheries management, one fisheries management agency representative explained that applicants for oyster lease licenses are screened with criminal record checks to ensure they are of sufficiently reputable character to enter the industry.

However in contrast, where there is political leverage in addressing the "social" in sustainable, fisheries management agency and industry organisation representatives demonstrate a greater understanding of the "social" in sustainable, and acknowledge there are a lack of social indicators and information on social issues in this more recently acknowledged component of sustainability. For example, in the rock lobster fishery at the time of the research, industry was attempting to generate sufficient data to respond to proposed management changes required under NCP. In this process, the fishing industry had to provide sufficient reason to prevent the fishery management introducing quota. Indeed, one industry organisation representative explained that the only reason social issues were being considered was to gain political leverage in an effort to stop the proposed changes and suggested that after the decision had been made, social issues would not remain on the agenda.

The University of Western Australia Centre for Regional Development were commissioned by the fishing industry and WA Fisheries to conduct a scoping study on the potential economic and social impacts of the proposed changes (Huddleston and Drew 2003). This research involved consultations with industry stakeholders. Women were not specifically targeted in these consultations and only mentioned in one recommendation, which suggested further information was required on employment opportunities for "fishermen's housewives" (Huddleston and Drew 2003, 46). This language and positioning reflects and perpetuates the secondary subject position of "fishermen's wives", reveals the association between this subject position and the appropriate place for women as being the "house" and highlights the lack of stakeholder status of women in fishing.

One fisheries management agency representative described the process of ascertaining data on social indicators in the rock lobster fishery as "challenging". However, it appears that much information could have been sourced directly from industry by fisheries management agencies through improved consultation processes. The fact that this does not occur suggests that the "knowledge" of women and men in fishing is not recognised as being as legitimate or valuable in comparison to the knowledge of qualified scientific experts such as researchers from the University of Western Sydney. Furthermore, outsourcing this research reflects the disintegration of the "social" in sustainable in fisheries management philosophy and practice.

Yet ironically, the extensive activities of many women in fishing in maintaining their businesses, families, communities and industry means they have significant knowledge and extensive potential to contribute to industry decision making, particularly on issues which are considered as the "social" in sustainable. This knowledge derives not from being "woman", but from the daily activities many women in fishing undertake as a result of the construction of gender, broader discourses of "wifing" and "mothering", and the subject position of "fisherman's wife".

However, it is unlikely that the knowledge of women in fishing will be recognised in the dominant masculine industry discourse where women lack stakeholder status, or in the dominant masculine fisheries management discourse where scientific, expert knowledge is privileged. In terms of the industry generally, acknowledging women as possessing this knowledge requires that their activities are legitimated as "work" and they are perceived not as "just fishermen's wives", but as industry contributors and stakeholders. In terms of fisheries management agencies, acknowledging women as possessing this knowledge would require recognition of the value of non scientific knowledge. Both of these strategies would require an expansion of "the industry" to incorporate women in fishing, which would reduce the power of men in fishing and fisheries management agencies. There is also a further significant implication of the current narrow construction of "the industry" as "fishermen" who engage in the singular act of harvesting in relation to the "social" in sustainable-the social impacts of fisheries management agency decisions *for* women, families and communities are not adequately considered. In the following section, I expose the subjugated knowledge of women in fishing by exploring women's understandings of "sustainability" and women's experiences of the social impacts of fisheries management decisions.

REVEALING A SUBJUGATED KNOWLEDGE

Women's Understanding of "Sustainability"

In contrast to fisheries management agencies, many women in fishing in this research conceptualised sustainability as equally incorporating social, economic and environmental issues, and in some cases, women in fishing prioritised social issues over these two other factors. Furthermore, discussions with many women reveal an expanded definition of "the fishing industry", where "the fishing industry" is constructed as more than the act of harvesting. Instead, many women refer to the industry as a "lifestyle", comprising people and families and communities, and where the succession of family fishing businesses is important.

> "…the **fishing industry is not about fish, it's actually about people**. The whole thing **revolves around people and people's livelihoods**... we're talking about people's lives and people who have a career in something that they love and enjoy. **It's a whole life, it's not a job** ... so when you change things you have a profound effect on peoples lives**, yet we take it all back to environmental and science reasons** without realising ... the industry is actually people" (woman, wild catch, in industry decision making)

Many women feel that social impacts are not adequately considered in fisheries management agency decision making.

> …**we could be managing the ... social impact a lot better** in some of our decision making…. It's something that's **lacking from a management perspective** ....a lot of people sit in Canberra and they **forget there's a face behind the names out there... a family** behind the name (woman, wild catch, in industry decision making)

There was diversity in women's capacity to discuss sustainability as a concept. Women in industry decision making often have tertiary qualifications and are experienced with industry discourses and, as such, are more likely to use language which currently prevalent in fisheries management discourse. Many of these women speak passionately about the need to integrate social issues into a more holistic decision making framework.

> "... **sustainable means the triple bottom line** thing and you've got to take that into account... with the Marine Parks ... some of the places actually lost school teachers because kids weren't there…the fishing industry moved. So therefore sustainability comes into infrastructure within a town. You're going back to…**emotional, mental and health effects .... sustainability is a holistic approach**. The whole way round, **it's just not the fish"** (woman, wild catch, in industry decision making)

Women in fishing who are not involved in decision making are less likely to discuss sustainability in a theoretical manner. However, these women also demonstrate a substantial understanding of the importance of the social in sustainability. The knowledge of the "social" for these women emerges not through their capacity to articulate the importance of social issues by using the language of discourses of "sustainability", but through stories of fisheries management agency decisions that have negatively impacted on themselves, their families, businesses and communities and others in the fishery.

Several women recognise that the meaning of "a sustainable fishery" is variable and contingent on an individual's history in the fishery, motivation to fish and financial situation. For example, smaller family fishing businesses may prioritise lifestyle, family business succession and the ecology and community in their local area, while larger fishing businesses may be more concerned with ensuring a stable or increasing profit margin.

> "....our fishery...has such a **wide range of people in it**...from very **small operators who for the third generation have been fishing the same grounds** and they say ...**"We've been fishing here for a hundred years and we're still catching fish and there's nothing wrong"**, to **very sophisticated company operations** and sustainability is very much to the fore of their calculations because **they're building big businesses on sustainability"** (woman, wild catch, in industry decision making)

Many women in fishing also feel that men and women have different understandings of sustainability. Several women also feel that, compared to men, women possess a greater awareness and a *different knowledge* of social issues.

> "Men are classic! **You talk to men about social impact**, they look at you funny. **'We don't know about this,** what should we know about this?' [laughter] …" (woman, wild catch, in industry decision making)

The majority of women in fishing with this view attribute this difference to "natural" embodied gendered qualities. However, several explanations also *alluded* to the knowledge derived from the experiences of women and men engaging in different daily activities. For example, one woman gives the example of women being aware of water shortages as they necessitate a change in the practice of washing clothes. However, the pervasiveness of the discourse of natural embodied gendered difference means that, despite acknowledging the gendered daily activities, this woman still attributes this difference to a "natural" difference in the way men and women think.

> "…it comes back to that **different way of thinking** again…. it's **definitely a woman's thing**. I **don't think men understand social impact at all** [laughter]. They know about economics and ... the environment but they

don't know about this social impact" (woman, wild catch, in industry decision making)

Several references were also made about social issues and impacts not "being seen". This correlates with the lack of visibility and recognition women in fishing receive for contributions such as unpaid work for the business and their activities in the spheres of family and community.

"In fishing we've got this ESD stuff that's supposed to be economic and social and environmental.... the **economic, you can see** figures... the **environment ....that's tangible too**, but **socials not tangible and they can't see that**... **Everything's really connected to the social side** if they just started to think about it, but I think **it's probably a women's thing**....we're the ones who are concerned if the school shuts down because ...half the fishers aren't there anymore, or if the doctor ... or the hospitals not there or the local post office doesn't work anymore or there's no childcare facilities cause there's only two kids left. **It's the women who have to worry about that"** (woman, wild catch, in industry decision making)

Social Impacts on Women, Families, Businesses and Communities

Exploring the social impacts of decisions on women, businesses, families and communities exposes women's subjugated knowledge. The most significant social impacts result from increasing fisheries management regulation and higher commercial costs which are altering the structure of the fishing industry. Smaller family fishing businesses are finding it increasingly difficult to survive and industry is becoming more corporatised. In order to remain competitive, family fishing businesses have to become more economically efficient and increase catches. This is achieved by investing in more advanced harvesting technology and larger boats. Industry is becoming increasingly "closed" with entry restricted to patrilineal inheritance practices or the investment of significant capital.

These pressures have resulted in a greater wealth divide between those who lease licenses and those with more secure ownership, and an industry culture where fishing is less of a lifestyle and more a "professional" business. There is evidence of increased competition between fishing businesses; increased stress on men, women and families due to increased debt levels and greater workloads in the context of unpredictable seasonal catches; an increasing number of small family businesses leaving the industry; and a reduction in the number and time of men and women in fishing available for industry leadership positions. Furthermore, as fishing effort increases, there is greater potential threat to fish stocks. This requires further regulation and restructuring of fisheries which, in some cases, necessitates the closure of fisheries or restructuring to reduce fishing effort through fishing license buy back schemes.

Changing industry culture

A cultural change is occurring in conjunction with the altered structure of the industry. Fishing is no longer a "lifestyle" and many women feel there is a loss of fishing community identity. Fishing is less a lifestyle and now more "a business" driven towards individualistic profit generation. Whether this transition is either necessary or beneficial in the quest for a more "sustainable" industry is questionable.

> **"Not [a community] so much** these days. Everybody is out for their own. It's **more of a business than a lifestyle** ....It's **not just fishing anymore**, its business and that's different.... The guys don't go fishing to go fishing anymore. It's too **money orientated"** (woman, rock lobster fishery, no legal ownership status, in her forties)

Increasing investment and pressure to catch has increased competition between fishing businesses, particularly for skippers who are employed or who lease licenses.

> "I don't like the pressure.... it's become **very, very competitive**....It's just **something totally different now. More of a family situation before**, but too many people own the pots outside the family, fishing companies and business people....It's...got away from the family situation. **It's more now an industry where those who are fittest survive**....it's just **cut throat**.... its not healthy" (woman, rock lobster fishery, European descent, family trust, sons who fish, in her sixties)

This competition is so intense that, in order to protect their family economic units, women in fishing are unable to discuss catches with other women irrespective of their friendship.

> "... we **don't talk about each other's catches ... It is competitive ....** It is just one of those **unwritten rules that you don't talk about that** ..." (woman, rock lobster fishery, two children, in her thirties)

The introduction of new technology is a further factor which has impacted significantly on the industry. In addition to increasing debt from investing in larger boats and more efficient harvesting gear, there are other implications from the increasing use of technology in the industry. Technology, such as mobile phones and laptop computers, means that women are able to communicate more regularly with their partners while they are at sea, reducing "worry" for women. Technology also means there is a greater potential for women to increase their business contributions and their "juggling" of responsibilities. One woman spoke about how improved access to technology on the Abrolhos Islands enabled her to conduct business dealings from this remote location, and an oyster farming woman explained how her laptop computer meant she could do bookwork at home during the evening rather than physically having to be at the oyster farm office.

Technology also means there is changing skill set required to harvest and fishermen no longer need to rely so much on generationally transferred skills. This potentially reduces the "natural fishing affinity" knowledge claim of fishermen and increases women's opportunities to engage in harvesting practices. However, it is unlikely that women's opportunities will increase as the construction of fishermen embodying innate knowledge appears discursively pervasive *irrespective* of the level of technological assistance fishermen employ in their harvesting practices. Indeed, the use of technology is associated with the emerging subject position "professional fishermen".

> "... **as technology grew**, I think **it's become more professional for the men**.." (woman, wild catch, in industry decision making)

Employing technology in harvesting and increased catches is also associated with negative public opinion.

> "... **as technologies got cleverer** we're using different techniques so therefore **we're catching more fish**, and that's where **the rape and pillage thing comes in"** (woman, wild catch, in industry decision making)

Several women feel that this perception, in conjunction with pressure from recreational fishers, is politically influencing fisheries management decisions including the closure of fisheries.

Increased pressure on family businesses

Many women described instances where smaller family businesses have been unable to compete with larger businesses and have been forced out of operation. In the context of the need to increase economic efficiency, smaller businesses are disadvantaged. One woman questions whether small family fishing businesses possess the financial and personal resources needed to increase their professionalism and efficiency.

> "... it's ... **very hard for people to do business compared to what it was** and **particularly at a grass roots level**, not so much an enjoyable industry. It was a lifestyle industry and now **people have had to run their businesses more professionally**, and while that might sound great **it may not be realistic** that they are able to. I mean if they could run a business before that was successful and they were happy with it and that was all they desired, then why should their business be any more professional if that comes at the price of them maybe having to get out of the industry**?....have people got the ability to ... compete with the big boys and are they going to be swallowed up with corporations?"** (woman, wild catch, in industry decision making)

In contrast to the wild catch sector where fish stocks are harvested, the aquaculture sector has greater resource control as fish stocks are farmed. In addition, businesses in this more recently established sector tend to be comprised of men and women with higher education levels than businesses in the wild catch sector. These factors are seen as significant for the future prospects of the aquaculture sector.

> "... **[aquaculture] has a brighter future**....they've **developed a vision** ...they've been really **good about drawing people into that process**. They've **got the skills**, they've got a more professional business base ... a lot of people in academia and ...they're very market orientated. They've got a resource that they have a little bit more control over.... **Wild fishing has a lot more challenges, a lot of people** in the industry **haven't got the necessary skills** ....They **need assistance**. They **don't have a vision** for the future. They've got a lot more ... environmental challenges, and **their futures not so bright**.."(woman, wild catch, in industry decision making)

Women are also under greater pressure as workloads increase. The complexity of bookwork and administrative requirements are significantly increasing and women are spending a greater amount of time on these activities.

"... very **different now** to what it was 25 years ago ...but again it gets back to the things that you have to do. I ... managed the house a lot better 20 years ago. … **now I battle to get the house done 'cause ... every day you seem to have to go the office.... the actual day to day running of the business involves a lot more..**". (woman, rock lobster fishery, in family trust, in her fifties)

For many women, taking on this greater administration role is associated with a greater sense of equality within their fishing partnership and compensates for their lack of involvement in harvesting. Several women also note that the increasing demands afford them the opportunity to become more involved in their business. However, many women experience stress from having to spend greater time and effort on these activities. In European fishing families, women make the additional contributions of interpreting and scribing.

As fishing capital in the form of quota becomes increasingly centralised, inequity increases between businesses who lease licenses and businesses with more secure ownership. The investment required to enter and remain in the industry also means younger women in the wild catch sector, and women in the oyster fishery who have bought their leases and are establishing their family businesses, have large amounts of debt, commonly up to or more than a million dollars. In addition to increasing administrative burdens, women in fishing in this research who lease licenses, younger women, and women in the oyster fishery are also often supplementing the family income while working harder in an unpaid capacity for their businesses and maintaining responsibility for households and caring for children. Many women describe stress as a factor in family and relationship breakdowns.

"…when you **place financial burdens on people and threaten people's livelihoods** ... there is a lot of **stress** and the impacts on the family unit ... are huge….**breakdowns in the family unit and marriages"** (woman, wild catch, in industry decision making)

Many women in this research in the rock lobster fishery felt that lifting the 150 pot rule would result in the accumulation of pots by larger businesses and the increasing expulsion of smaller family fishing businesses.

"The 150 pot rule…. I don't think it should be taken away because **it still gives the smaller boats a chance**. If you are a big boat fisherman you would want that to go because you need more pots to be able to pay for

> your big boat…" (woman, rock lobster fishery, in her thirties, three children, in partnership, in her thirties)

The majority of older women with a long standing industry involvement, younger women in smaller family businesses and women involved in businesses that did not own all their pots, also feel that going to quota management will be detrimental to the industry and did not want these changes to occur. Currently in the rock lobster fishery, seasonal restrictions combined with unrestricted catches mean that fisher(men) harvest almost every day during the eight month long season. During this time, the majority of women take on the full responsibility for managing the family and household and many aspects of the business. Going to quota has the potential to reduce this burden on women. While some women recognise this, many women are concerned about the additional social impacts of going to quota.

Women felt that these additional social impacts of the proposed change could include: escalating costs and additional administration requirements; increasing stress from operating in an increasingly competitive environment; smaller, less financially secure family fishing businesses exiting the fishery; and the loss of small, coastal communities and a further loss of identity within the fishing industry.

> Participant 1:"There will be **more paperwork. Bloody hell! We'll have to pay** for x amount of fishery inspectors to be standing on the jetty to check all your paperwork ... "
>
> Participant 4: "It is going to be **a lot more expensive"**
>
> Participant 2:" Yes and **it all comes from fishermen** so **it's going to cost us.** It's no benefit to us and then you are going to have fishermen who's got **paperwork** and he might miss putting in that calculation **and it will be wrong and he will get fined"** (rock lobster fishery, focus group)

Women felt that going to quota would also restrict the catches of larger boats leading to dishonesty and further regulation, and perpetuate the need for further compliance measures, greater costs and administration. The nature of fishing capital also means that in a regulated environment it is difficult to sell fishing assets.

> "If you bring in quota there will be **leakage and more dishonesty**. People will start selling quotas underhandedly….most of the big boats at the moment have 100 –123 pots in the water …and all of a sudden they are

given this restricted quota. They **wouldn't survive....** **where would they sell the big boats?"** (woman, rock lobster fishery, family trust, sons who fish, in her sixties)

Escalating costs to operate in fisheries force fisher(men) to increase their fishing effort to survive. These increasing costs are a consequence of fisheries management agency cost recovery policies and investments in larger boats and more efficient harvesting gear. Thus, an offshoot of this social restructuring of the fishery also has potential environmental consequences. Many women feel the cycle of greater fishing effort to service fisheries management costs and fishing capital debt means that stock levels are increasingly threatened. Again, while economic efficiency and sustainability are conflated in management discourse, this raises the question of how necessary and beneficial current fisheries management interventions are in the quest for more "sustainable" fisheries.

"... that's the way it is.... **bigger the boat, the more you have to catch"** (rock lobster woman, in her thirties, three children, in partnership)

Some women believed that, in contrast to smaller family fishing businesses who are concerned to preserve their environment in order to maintain their lifestyle, livelihood and pass their business down, larger fishing businesses pose a greater threat to fish stocks and the environment, as their primary aim is to maximise profits. This is the concern of one woman who describes the process where, under NCP changes, licenses will be available to larger fishing businesses.

"Up until now **fisheries like ours are owner-operators... that's what keeps it sustainable** and ...the environment in good condition ... an owner-operator has to look after it.... competition policy is probably going to be the end of it... **companies will be able to buy out these licenses and...that will ... be threatening to the fish and... environment** ... it could be quite devastating ....Because people that do that sort of thing usually don't care too much.... They just come in and catch their fish and go out again.... a little bit like the fly in fly out mining...[they] don't contribute anything to the town. Their families are back in another town.... **they just come in, fish, and go back to where they belong, and they don't care very much then whether ...the biodiversity is intact. It's not their worry"** (woman, wild catch, in industry decision making)

Exiting the industry

Many women in family businesses also feel uncertain about their future, and disempowered in the face of management's power to dictate regulations which could impact significantly on their lives. Women speak about the impacts on small coastal communities as smaller family fishing businesses are forced to exit the industry and relocate to seek alternate employment.

> "It is all about **small communities being wiped out**....unfortunately it's the small families and the small business people that fall by the way side" (woman, wild catch, in industry decision making)

While restructure schemes often include provisions for (fisher)men such as retraining and assistance in seeking alternate employment, the fact that the work of a "fisherman's wife" is unpaid and unrecognised means that even where women provide extensive labour to their enterprises, women may be not offered these opportunities.

The following description by a woman, who discusses restructuring in the Tasmanian rock lobster fishery through the implementation of quota management, exposes the agenda of fisheries management to displace smaller family fishing businesses.

> "... husband, wife and children operations ... all regional community people ...in terms of the flow on effects into towns....the government ... wanted to reduce the amount of effort in the industry and get people out, so they put a restriction on how much fish could be caught ...how many boats... licenses and ... pots in the industry.... **the government truly believed that the industry should be corporatised ....as in 'get rid of the smaller operators and then we'll only have bigger operators'** and as far as I'm concerned that's arse up....Tasmania is about niche markets ... and we should be exploiting that ...and **the state government ...seemed to think that '...we need to get up with the big boys and get rid of this cottage industry mentality'**....you've lost this whole identity in the industry.... this **tradition** and **culture**.... **you've lost three, four generations of people doing this"** (woman, wild catch, in industry decision making)

This woman describes the social impacts of restructuring within this fishery. These include difficulties in complying with complex administrative requirements in smaller fishing businesses where errors are heavily penalised; smaller businesses being forced to either invest in further capital or exit the industry; and difficulties in finding alternative employment for older (fisher)men. Presumably, many women partners of these (fisher)men also experience difficulty in locating paid employment.

> "sometimes they'll…write in the date as numbers ... and that's invalid … the department will...say '.... you're going to have to fill in a statuary declaration... and you're not allowed to leave port until you've fixed it'.... **if they're in their fifties and sixties … it's really insulting for them to have someone dictate things over what they consider to be quite petty. In terms of managing that change,** nothing was actually done. It **was just implemented and a stick was wielded**....
>
> But the people who have had to get out of the industry, **at that age realistically what are they going to do?... it's all they're ever done.** They started on the boat when they were 16 with their dad.... nobody wants them as a deckhand at that age. **They don't have the skills to start over and they're at an age where they can't start over**..".(woman, wild catch, in industry decision making)

Some fisheries have been restructured to reduce fishing effort and the numbers of licenses through fishing license and capital buy back schemes. One woman discusses the social impacts of the government buy back scheme in Queensland that was implemented as a result of the Great Barrier Reef Marine Park rezoning, which took effect from 1 July 2004.

> "... they got rid of a third of the trawlers....the buy back scheme in Queensland has been **dreadful ... fishermen haven't been able to leave the industry with any dignity**. The buy back scheme created **suicide and lots of problems**... it was very **traumatic**.... we had no job, no nothing. And that was it, **no say, no nothing"** (woman, wild catch, in industry decision making)

These social impacts of the closure of large parts of the Great Barrier Marine Park to commercial fishing businesses are discussed in a television program discussion (*Fishing crisis* 2006), in which Dr Daryl McPhee states:

> "The real cost in economic terms hundreds of millions of dollars. The **real cost in social terms unquantifiable**…**I've had the task of talking commercial fishermen out of suicide** when I've been doing my studies because they're that **stressed, they are that concerned, they have nowhere to turn**…Increase in **suicide rate, increase in domestic violence** – …**negative social impacts that arise from dislocating families who only know one thing and that's fishing"** (Fisheries crisis 2006)

"Closed entry" of fisheries

Increasing costs and regulation is resulting in the industry becoming increasingly "closed", with entry restricted either through patrilineal inheritance or the investment of significant capital. The opportunities for young people with no prior family involvement to enter the industry are diminishing. In the context where licenses are often not legally able to be split or, as in farming, where the splitting of licenses is not possible as it would render the business economically unviable, patrilineal inheritance practices are reinforced as a mechanism for children (sons) to enter into the industry, perpetuating the exclusion of women (daughters).

> "... it's not an industry where you want to pull out of very quickly. It's too hard to get back into it. It is **very difficult to get a loan and get established** .... I've got **3 boys ... one of them will eventually take it on....without family involvement there is practically no way that young kids can get into the industry** on their own these days. You need 3 million dollars to start with just to get a boat and licence" (woman, rock lobster fishery, in partnership, three sons and one daughter)

It can also be suggested that the increasing difficulty in entering the industry exacerbates the reluctance of mature women, particularly in ethnic families, to pass control of bookwork and their businesses to their sons and daughters-in-law in the event they divorce. While this disadvantages women and operates to reinforce their lack of stakeholder status, many daughters-in-law appreciate the financial opportunity provided through family businesses, which makes it possible for their partners and themselves to operate in the industry. In contrast, in the oyster fishery, while large capital investments are also required to enter the fishery, due to the lack of tradition of patrilineal inheritance practices, it is more common for families to go into significant debt and for women to contribute substantial unpaid labour to their enterprises.

The reduction in the number of fishing businesses in the industry also means there will be less young people, and fewer women and men in fishing within the industry overall who are available for leadership roles. The opportunities for women to engage in manual labour and harvesting practices are also reduced as the overall number of women in fishing declines with fewer numbers of fishing businesses. As harvesting becomes constructed as an increasingly "professional" occupation, increasing efficiency and capital investment in businesses means deckhands are also likely to appropriate the manual and practical contributions of women in fishing businesses which are financially secure. However, where licenses are leased and businesses are under threat of survival and, in the oyster fishery, where families were often in substantial debt, women still contribute significant manual labour in their businesses.

Formalising women's marginalisation in resource ownership

The formalisation of the allocation of fishing rights reflects the centralisation of those rights according to capital ownership-an area where women are marginalised (Munk-Madsen 1998). Fundamental to the manner in which these resources are allocated is the concept of production which relates to harvested fish tonnage and equates directly with economic remuneration. Women's industry contributions are excluded from this concept of production and women's lack of stakeholder status excludes them from consideration in the process of the formal allocation of the rights to resources.

Privatising the rights to resources compounds women's industry marginalisation by formalising previously informal patriarchal relations. Munk-Madsen (1998) describes how this applies to women who harvest fish. Based on this doctoral research, I have argued elsewhere (Dowling 2006a), that women in family fishing businesses are also marginalised in this process given the prevalence of patrilineal inheritance in the industry.

In the rock lobster fishery, the decision to go to quota management was expected to be made in 2007. Fisheries management recognised that the allocation of quota-"*the best management system*"-would be a "contentious and difficult process" and that one *"cost"* of this process would be the marginalisation of some fishing businesses. Potential social impacts of this change are absent in his discussion.

> "…that's **one of the costs** ... not only in time and effort, potentially in compensation but **in terms of the politics**…because people have ...large investments….I **haven't put a lot of thought into how the allocation process would work** yet... It would be I **expect a somewhat contentious and difficult process** but ... I don't think it should be something that shies us away from going down the path **of what is the best management system**..".(male representative, management, rock lobster fishery)

Women's lack of ownership and stakeholder status mean that women in fishing will be marginalised in this formalising of the rights to resources in the rock lobster fishery. However, in the contentious process of restructuring, women are likely to collude with the dominant discourse and unite with and support their partners to ensure the best possible outcome for their family economic units.

Furthermore, women in fishing feel that consultation processes for the proposed changes have been inadequate and that the views of industry have not been heard. Given women's lack of stakeholder status and the lack of management consultation processes, it seems unlikely that women's knowledge of social issues will be given consideration by management and industry in their current endeavour to gather information on the potential impacts of these changes.

The Potential of Women's Subjugated Knowledge

Women in fishing have a significant contribution to make to fisheries management and the future of the fishing industry by contributing their knowledge of social issues gained through their daily activities. Indeed, several women in this research feel that women's increased representation in industry and management decision making where they could contribute this knowledge would result in different and more beneficial outcomes for industry.

> "with the ESD ... **its important that women are involved** because if we're talking about reducing numbers, if we're talking about reducing gear, if we're talking about environmental factors, **women should be at the board table.... because ... they will bring a different perspective** ... to ensure that when those policies are designed, it will come from a **holistic approach"** (woman wild catch, in industry decision making)

Increasing recognition of the social component in sustainability by some fisheries management agencies, in conjunction with the acknowledgement that fisheries management lacks social data, provides a potential opportunity for women to contribute their knowledge.

> "...**we [management] recognise that it [social impact] is an area where we are deficient** from an information perspective ...I can't say what's the right way of doing it because we just don't know....**so let's find out so we can be informed** on how we go forward..." (male representative, fisheries management agency, rock lobster fishery)

Fisheries management agencies fail to take up the opportunity to access this knowledge through consultation, despite WINSC providing an avenue for fisheries management agency to both access information from and disseminate information to women (and men) in fishing.

> "... **governments, or whoever wants to be able to talk to a group of industry**...it would be **really valuable** for them to be able to **access women** in the industry.... instantly the **right people to talk to** ... because these women are then taking that message or that question back .... you are talking about an **influence in the industry**... **a catalyst to getting the information disseminated** to the broader community or the men in the industry..." (woman wild catch, in industry decision making)

Women in fishing have the potential to demonstrate resistance, assert their knowledge claims and bring social issues to the fore. Women in the rock lobster fishery who lobbied government successfully and overturned the home porting rule (see the section Collectives of women and alternate discourses in Chapter 5), collectively asserted their knowledge of social issues and the social impacts of management decisions.

> "...the best thing we did was took some of our **fishing families with us, the kids came and the wives came**, and that ... was **our way of saying that our industry is about families and not about the men**, the fishermen" (woman wild catch, in industry decision making)

The fact that this decision was overturned indicates that the social implications had originally not been adequately considered, and that the knowledge women in fishing brought to the debate was knowledge not previously considered. It also demonstrates the potential value for fisheries management agency and industry of recognising and accessing women's knowledge *prior* to decisions being made.

WINSC has also displayed activism in relation to asserting social knowledge claims. Indeed, WINSC actively promotes "the industry" as being *more* than fishermen.

> "... WIN either at a state or a national level has ... to ensure that **women of the industry are up there**. That **when government think about policy**, they're not only thinking about the industry, **they also are aware that there are women** involved, and **families** involved and **communities** involved. It isn't just about taking a fisher out of the water. It has that **whole impact across the board"** (woman wild catch, in industry decision making)

In response to the political closure of a fishery in South Australia and in recognition that further fisheries are under political threat, WINSC developed a draft social impact plan for government to ensure that there is a process which can be followed when fisheries are being restructured. WINSC developed this plan by talking with industry, and particularly with women, about the social impacts of fisheries closures.

> "Currently there is **no chart or guide that a government can follow in regard to a restructure** of a fishery. So we're working on that **social impact stuff because it is the women who feel the brunt of the issues**.... the woman is the person who **pulls the family together**, and the **family revolves around the business**, and the **business revolves around the small community**. So women are very good at trying to nut those issues out, work through those issues" (woman, wild catch, in industry decision making)

This social impact assessment guide was published by the BRS (Schirmer and Casey 2005). However, this publication does not acknowledge that the project was initiated by WINSC, or that WINSC provided extensive data and support to the development of this document. Indeed, WINSC did not seek this recognition and instead, intentionally adopted a strategy where WINSC ensured government received the accolades for this document. This colludes with the dominant discourse whereby women in fishing are not recognised as stakeholders and have no legitimate claim to knowledge. However, this strategy was intentionally employed by WINSC to facilitate the likelihood of their objectives being achieved with the least level of resistance from fisheries management agencies and industry.

> "... we recognised this two years ago, so **it's been on our agenda, but not highlighted** ... because ... **we want to see the government as raising this and doing this**.... **We don't take the accolades**, but we put the suggestions there, provide the information and do all the tough stuff and **let them take the credit"** (woman, wild catch, in industry decision making)

While there is potential for the knowledge of women in fishing to benefit the future of the industry, unless fisheries management agencies and industry recognise women's knowledge, and facilitate their involvement by addressing the discursive structural barriers to women's greater involvement, it is unlikely that women will gain significant voice. Encouraging women in fishing to share their collective knowledge is empowering and may have some influence. However, this alone is an insufficient strategy as it attributes responsibility for women's greater involvement to women, and potentially absolves fisheries management agencies and the broader industry of the responsibility to address discursive mechanisms that exclude women.

In addition, recognising women as stakeholders and legitimating their knowledge in the current pervasive masculine industry discourse threatens masculine power as it necessarily requires an expanded reconstruction of the industry as more than harvesting and more than fishermen. In order to enable women in fishing to make a legitimate knowledge claim, it also requires women's activities to be acknowledged as more than the activities of "wife" and instead as vital contributions to the industry. Women may be reluctant to demonstrate resistance in the current climate of fisheries in crisis given the potential costs to their personal relationships and their family economic units.

A further issue which emerges from exploring the discourse of sustainability in the industry is the need for improved consultation processes that enable the voice and knowledge of "grass roots" industry, both men and women in fishing, to be incorporated into decision making processes. Enhancing industry sustainability will be illusive until this occurs and all stakeholders are included in decision making processes. Fisheries management agencies are charged with the responsibility for ensuring a "sustainable industry, and as such, they have a responsibility to ensure that funding is dedicated to implementing strategies which include all stakeholders. This is particularly important given the barriers for women to engage in decision making at higher levels.

Conclusion

The current discourse of sustainability in the Australian commercial fishing industry espouses the integration of social, economic and environmental issues. However, fisheries management is based on a bio-economic model which prioritises environmental and economic factors. Deconstructing the discourse of sustainability in the context of crisis reveals an agenda to displace smaller family fishing businesses.

Fisheries management agencies demonstrate a lack of awareness about social issues and the social impacts of fisheries management decisions. Indeed, some of the social impacts of fisheries management decisions which have been made have resulted in dire consequences for many men and women in fishing. While social issues are considered in particular circumstances, such as in the rock lobster fishery at the time of this research, this consideration appeared to be generated to gain political leverage and, as one industry organisation representative commented, after this time has passed, social issues are unlikely to remain on the agenda. In contrast, where there is no leverage in addressing the "social" in sustainable, such as in the oyster fishery, fisheries management agency and industry organisation representatives demonstrate a significant lack of awareness about these issues.

The announcement by the Commonwealth Minister for Fisheries in late 2006 of a buy back scheme in all Commonwealth fisheries meant significant changes for women and men in fishing in Cardea. Given the increasingly unviable status of their fishing businesses, they may have no options than to accept inadequate compensation. Additionally, families who are forced to exit the industry have limited skills for alternate employment and live in a small coastal town with a high unemployment rate.

While "knowledge" on social issues is being sought in some fisheries such as the rock lobster fishery at the time of this research, women's invisibility and their lack of a legitimate claim to knowledge mean they are not recognised as stakeholders in debates on industry sustainability or as possessing knowledge. Women's lack of visibility and stakeholder status mean firstly, that women are excluded from being considered in the process of the allocation of the rights to resources and secondly, that the social impacts of management decisions for families, women and communities are not adequately considered.

In this conflict between "us" ("fishermen") and "them" (fisheries management), both parties claim superior "knowledge". "Fishermen" claim legitimate knowledge (power) based on experience and the "knowing" of fishing, while fisheries management claims legitimate knowledge (power) based on educational qualifications and discourses of positivist Truth in science. In the context of this conflict, many women in fishing resist fisheries management by colluding with the dominant discourse and supporting their husbands in order to preserve their family economic units. This conflict reduces the discursive space available to women to assert their own claim to knowledge, as doing so potentially detracts from a unified industry resistance. There is no space for women in the "us" and "them" struggle over the privileging of knowledge between fishermen and fisheries management.

Discussions with women about sustainability and social impacts reveal a subjugated discourse which acknowledges the "social" in sustainable. Women demonstrate an understanding of sustainability which incorporates lifestyles, families and communities. This knowledge derives from the experience of many women who engage (invisibly) in specific discursive practices of "wifing" and "mothering". Ironically, it is this same (invisible) collusion which both enables women to gain knowledge, which also means women have no legitimate basis for claiming knowledge (power) and positions women as lacking stakeholder status.

While some women are demonstrating activism and asserting their knowledge by contributing to the discourse of sustainability in the industry, women are not recognised as possessing knowledge. Inadequate consultation processes ensure that women's knowledge remains untapped. Addressing women's lack of industry visibility and providing women with recognition to legitimate their substantial and unique knowledge is vital for addressing existing inequalities, preventing the further entrenchment of this inequality and for enhancing the future of the industry. Deconstructing the dominant fisheries management discourse and exploring the "view from below" (Harding 1987) exposes the subjugated knowledge of women in fishing. This knowledge has the potential to form the basis for a standpoint for women in fishing.

CONCLUSION

Based on my research and the evidence presented in this book I conclude that the Australian commercial fishing industry is locally, politically, systemically and structurally gendered. Despite making significant contributions to the industry, their businesses, families and communities, women in fishing lack visibility and stakeholder status. While some women in fishing demonstrate agency, resistance and activism, women in fishing are fundamentally marginalised through a range of mechanisms and barriers. Furthermore, in the current climate of increasing legislation and bio economic, masculine, fisheries management intervention in a neo liberal state, it is likely that women in family fishing businesses will become increasingly marginalised in the quest for more "sustainable" fisheries.

A Note on Theory

Prior to overviewing my argument and the critical findings of this research, there is a need to reaffirm my theoretical position on plurality/ essentialism so that I may speak of "women in fishing".

As previous research has suggested, there is a need to go beyond documenting women's "roles" to explore power relations (Davis and Nadel-Klein 1992), how women are positioned in industry discourses and the effects of this positioning (Liepins 1996). How power is exercised through the industry and fisheries management discourses is central to understanding gendered industry power relations, the marginalisation of women in fishing and the opportunities for their resistance. Employing a post structural framework expands the potential to theorise the possibilities for women's agency in addition to recognising the constitutive force of discourse and how social structures emerge through these discourses.

Language constitutes and is constitutive of subjectivities, and shapes the way subjectivities and social structures are constructed, challenged and defined (Foucault 1980). Meanings are not guaranteed by the author/ speaker but are inherently unstable and constructed within specific cultural and historical contexts. Not only is conceptualising language in this manner critical for my deconstruction of the concept of "sustainability", but it also requires that women's lack of visibility and their contributions are analysed as *located within* the changing nature of the industry and shifting broader discourses.

While post structuralism recognises plurality and diversity, reconciling post structuralism and standpoint theory enables generalisations to be made on the basis of knowledge being located in similar, shared material experiences such as labouring. While I acknowledge and remain committed to recognising difference "within", women's shared experience (or at least aspects of shared experience) means threads may be drawn together to speak of the similarities of experience and knowledge of "many" women in fishing. In doing so, I acknowledge my imposition (based on my partial and situated knowledge) of a certain fixity onto those occupying the social positions of "woman" and "ethnic". However, I feel that this "slippage" towards essentialism (Wilkinson 1998a) is necessary on political grounds and, the shared experiences of members of a group do not mean that all individuals share an identical experience (Collins 2004b; Hartsock 2004b).

Furthermore, in this instance, there can be no ambiguity about the *pervasively masculine nature* of the dominant discourse and the discursive division of labour which correlates so strongly with the subject positions of "fisherman" and "fisherman's wife". I argue that the significant but unrecognised knowledge many women in fishing gain derives not from "woman", but from difference, that is, undertaking their daily activities (Haraway 1988).

An Overview of My Argument

This book provides evidence that the Australian commercial fishing industry is characterised by a dominant masculine discourse, which narrowly constructs the industry as comprised of (fisher)men who practice the singular act of harvesting. This masculine discourse accounts for the pervasiveness of women's lack of visibility, the lack of recognition for their contributions, their lack of stakeholder status and their poor representation in industry decision making. (Fisher)men are privileged with occupational and stakeholder status and women are relegated to/ take on the secondary subject position of "fisherman's wife". However, despite their lack of recognition, women in fishing make vital contributions to maintaining their industry, businesses, families and communities. A rigid discursive division of labour is a regulatory mechanism of the dominant industry discourse which normalises certain activities such as harvesting for those occupying the subject position of "fisherman", and "doing whatever needs to be done" for those occupying the subject position of "fisherman's wife". The normalising function of discourse means that these activities are constructed as "natural" and those who breach these discursive constructions are viewed as abnormal. Women do not separate their contributions into discrete spheres of industry, business, family and community but experience their daily activities in a more holistic manner as "lifestyle". I argue that the social construction of gender means women are more likely to engage in multifaceted social/ community issues, and that as a result of engaging in these activities, women may make a knowledge claim.

The current industry climate is characterised by increasing regulation and discourses of sustainability in response to "crisis". However, while industry discourses of sustainability espouse the importance of the integration of environmental, economic and social factors, bio-economic fisheries management fails to integrate these factors or adequately consider the social dimension. Some fisheries management agencies recognise there is a knowledge gap relating to social issues relevant to sustainability. Deconstructing language and the meanings associated with language reveals the political and masculine nature of a seemingly apolitical fisheries management system. Women in fishing are not recognised as stakeholders and the social in sustainable is secondary in relation to (masculine scientific) environmental and economic factors. The implications of women's lack of visibility and stakeholder status means that the social impacts of management decisions are not adequately considered and that women are further marginalised in the allocation of the rights to resources.

Furthermore, state legislation underpinning management intervention provides fisheries management with the power/ knowledge and right to claim Truth-over "fishermen". Through language and the manner in which meanings are conveyed by fisheries management, smaller family fishing businesses are being excluded from the industry. Women in family fishing businesses support their partners in this struggle against management. In this context, the potential for women to claim discursive space where they may assert their knowledge and stakeholder status is significantly reduced as doing so would detract from a collective industry position. Furthermore, regulatory discursive mechanisms function to exclude women from the practice of fishing and so prevent women from claiming occupational and industry stakeholder status; ensure that the construction of women's activities as "natural" for "fishermen's wives" means that women do not require recognition; penalise women who breach the dominant discourse by constructing them as "abnormal"; and exclude women from industry decision making.

I argue that in the quest for more sustainable fisheries, women are becoming increasingly marginalised and gender inequity is likely to become further entrenched. Women in family businesses are already marginalised within the more traditional narrow construction of industry as comprised of "fishermen" who harvest. Women are further marginalised by the exclusion of smaller family fishing businesses occurring through the current re/construction of industry driven by neo liberalist economic rationalism that is disguised in the quest for sustainable fisheries rhetoric.

I contend that more equitable gendered industry relations are unlikely without a specific focus on increasing women's recognition as stakeholders and recognition for their activities at a structural level. However, this would require management and industry to value and recognise women's contributions and women's knowledge to be legitimated. This would involve women being acknowledged as industry stakeholders and the reconstruction of industry as *more* than fishermen and as *more* than harvesting. Having demonstrated that the current industry and fisheries management discourses are masculine, this is unlikely to occur, particularly as the emerging subject position of "professional fishermen" is displacing "fishermen" and their families.

However, women do demonstrate resistance and have significant potential to enhance the future of the industry. I agree that the industry is in crisis and this situation is likely to be exacerbated until an integrated discourse of sustainability emerges which is not driven by economic rationalist agendas and which incorporates governance where all stakeholders have voice, and where knowledge is recognised and valued. The opportunities for women to have voice in this manner requires a industry/ fisheries management discourse where market forces are not employed to achieve sustainability, that is, privatising the rights to resources, but where social justice and inclusion are incorporated into the quest for sustainable fisheries at more than a level of rhetoric.

Critical Findings

Dominant masculine industry discourse

I conclude that the Australian commercial fishing industry is pervasively masculine and locally, politically, systemically and structurally gendered. Deconstructing language and exposing meaning associated with the apparently apolitical concept of the "fishing industry" reveals a common understanding that "the industry" is comprised of (fisher)men who (are privileged to) engage in the singular task of harvesting fish. The remarkable similarities across the three case studies demonstrate the pervasiveness of this construction. Women's substantial industry contributions in a range of other areas are not recognised and women lack stakeholder status.

The Western dualistic and relational nature of concepts is evident in the industry construction of fisherman/ masculine/ man and fisherman's wife/ feminine/ woman. These dualistic dichotomies are also hierarchical and imbued with power relations. The term "fishermen" privileges men with occupational status. Women in fishing lack occupational status and instead are defined by marital status, and relegated to/ adopt the subject position of "fishermen's wife". The secondary positioning of women in fishing is also revealed in further exploration of language. Many women in fishing speak about being *"behind the fisherman"*, "in the *background"*, and *not being "seen"* or *"heard"*. Furthermore, the pervasive use of possessive language inherent in descriptions of women in fishing-for example, the term "fisherman's wife" may be read as the "wife belonging to the fisherman"-is not used when referring to men in fishing. Constructing women on the basis of marital status reflects and perpetuates women's secondary positioning.

The masculine occupational privileging of harvesting is reflected and perpetuated through discursive regulatory mechanisms which operate to exclude women and ensure they do not appropriate the masculine domain of harvesting. In the wild catch sector, women are excluded from accessing masculine discursive sites of power-boats and wharves-and the "natural" and "blokey" qualities associated with the subject position of "fisherman".

However, many women in this research do fish or have fished. When women resist the construction of harvesting as a masculine occupation by asserting their own skills, knowledge and right to fish, they are derogatorily constructed as "abnormal"-as being a "certain type", "like a man", "butch", "blokey", or as sexually "loose". In contrast to men, women are further disadvantaged when they develop muscular bodies as a result of fishing. Women are excluded from fishing on the assertion that they lack the physical capacity required to fish. However, common sense theorising suggests that fisher(men) already possess a certain physical capacity *as this is a prerequisite to fish*, whereas a similar level of strength is not required to marry a fisherman. Furthermore, there is variability in the strength amongst men, and presumably some men too lack the physical capacity to fish. These associations between femininity and physical bodies act to reproduce sexist discourse, equate the feminine with "woman" and so reinforce the distinction between "women" and "men" (Gough and Edwards 1998).

The normalising function of discourse that ascribes the "naturalness" of men fishing and the "abnormality" of women fishing is a powerful regulatory mechanism. The construction of "fishermen" as "men" means that no matter what women do, they cannot ever achieve the status of fishermen, as they are "women". Women are excluded purely on the basis of the gendered construction of bodies. Many women in fishing collude with this discourse, perpetuating gendered power relations by suggesting that fishing is an undesirable occupation for women, including their daughters. This also demonstrates how meaning is transferred from generation to generation.

Patrilineal inheritance practices reflect and perpetuate masculine industry privileging and constrain the entry of women into the industry through marriage. While patrilineal inheritance is prevalent in both Cardea and the rock lobster fishery, many women are involved in collective business structures. This provides some evidence to support the assertion that there is a trend in Australia towards transferring family farming businesses through collective business structures such as trusts and companies (McAllister and Geno 2004). These collective structures have the potential to foster more equitable gendered industry relations. However, in this research, where family fishing businesses adopt a trust or company structure, this is often to achieve optimal taxation benefits for businesses, and several women with legal status in collective structures refer to the business as "being" or "belonging to" their husbands. Thus, it can be suggested that despite some evidence which supports this trend towards collective business structures, inequitable gendered relations are likely to remain while men are occupationally privileged. Indeed, this privileging of men within seemingly more gender equitable legal arrangements may mean the marginalisation of women in fishing is more subtle and insidious.

Many women in this research adopt/ are relegated to the subject position of "fisherman's wife". However, acknowledging diversity amongst and within women in case studies, the degree to which women take up this subject position differs. Women who have been exposed to alternate discourses of subjugated knowledge about gender equality through education or experience are more likely to have subjectivities located within multiple discursive scripts and resist their construction as "*just a fisherman's wife*". However, reflecting the integration of sex and ethnicity, women in Cardea, who are all involved in European family fishing businesses, are more likely to rigorously adopt and adhere to the subject position of "fisherman's wife". The more rigid discourses of "wifing" and "mothering" in this case study site may be attributable to these families forming a closed community and working together collectively in order to survive with limited economic and social power after immigrating to Australia. This also accounts for the pervasiveness of patrilineal inheritance practices which are evident in European fishing families.

There is an inherent complexity in the manner in which language and meaning intersect with differing women's subjectivities in the exercising of power to construct/ reconstruct other women as abnormal. This is evident in the derogatory constructions *between* women who collude and women who resist. Acknowledging this diversity within the subject position of "fisherman's wife" reflects the diversity amongst women. For example, some women in fishing who grew up in non urban areas and who married local fishermen, and women who have fathers who are fishermen, ostracise women who do not adhere as rigorously to the subject position of "fisherman's wife". In turn some of these latter women–typically women with tertiary education, who "choose" to work and/ or buy into the industry–construct women who more rigorously take up the subject position of "fisherman's wife" as "insular" and/ or "just a wife". This was also evident in the re/construction of women who are and who are not involved in industry decision making. Thus, derogatory constructions by both women who collude and resist may be seen as perpetuating the dominant discourse.

The implications of men's occupational privileging cannot be underestimated. Men "belong in" and "own" the fishing industry. Women in fishing are marginalised in the ownership of fishing capital and businesses, and are excluded from decision making.

Discursive division of labour

As in fisheries overseas, women in the Australian commercial fishing industry make vital contributions to their industry, businesses, families and communities, yet these contributions are not recognised. Particular tasks are discursively normalised and "natural" for certain subject positions. "Fishermen" fish and "fishermen's wives" have the "job of wife". This involves nurturing and caring for family and community, supporting the business and industry, and doing "whatever needs to be done". However, women's secondary positioning is evident as their labour and activities are not constructed as "work", but instead are viewed as activities "naturally" undertaken by "wives" ("women"). While individual partners of some women may recognise what their partners do, at an industry level there is no acknowledgement of the contributions of women.

This deconstruction of the "natural" and "normal" tasks undertaken by "fishermen" and "fishermen's wives" reflects the relational and hierarchal dualisms inherent in gendered discourse. "Men's work" is hard, dirty, rough, physical, masculine and *privileged* over "women's work" which is caring, nurturing, clean and feminine. The alleged "natural" difference between men and women justifies the distinction, or the "fair division of labour", between the work of men in fishing (harvesting) and women in fishing (numerous other tasks). The unified nature of the economic family unit further reinforces the division of labour where it is necessary for family members to work together across the spheres of industry, business, family and community.

This commonly accepted division of labour, where it is normal for men to fish and for women to undertake the activities of "wife", may be seen as a discursive regulatory mechanism, which operates to segregate tasks based on the construction of women and men. The *discursive nature* of the division of labour (in contrast to the division of labour being constructed as a set of rigid social relations which bindingly and irrevocably assigns women and men to particular roles in an explanatory functional theoretical model) is also evident in that it is not absolute. In this I refer to the fact that women in fishing may engage in activities usually taken up by or relegated to men such as harvesting fish, *without* this engagement constituting a discursive breach. This occurs where women undertake these activities to support and "help" their partners and where women do not claim recognition for their contributions. This is particularly acceptable where women do this in order to support their families and businesses in times of financial hardship or fill in for an absent crew member. In these instances, this does not constitute a breach as women remain primarily located within the subject position of fisherman's wife-secondary, supporting and invisible. However, women who resist engaging in the activities associated with the subject position of "fisherman's wife", or women who seek acknowledgement for these activities, are derogatorily constructed by many men and women in fishing. In a similar manner to the re/construction of women who fish in their own right, women are punished for breaching the dominant discourse and constructed in a derogatory manner as "not good wives", "different" and/ or "selfish".

Demonstrating the self regulating power of discourse, women's collusion may at least partially be attributable to the productive nature of power which affirms identity (Foucault 1980). Women who engage (quietly) in the activities associated with the subject position of "fisherman's wife" are not only able to claim this identity as their own, but are able to identify and be identified as a "*good*" wife. However, many women speak about "loving" what they do, which may also provide support for the Foucaldian feminist critique that emotions and internal sources of empowerment must also be considered in relation to agency (Deveaux 1994). A further factor in women's collusion may also be the integrated nature of the economic and family unit, where women may exercise power within their private home sphere and where the penalty for breaching discursive practice may result in the personal cost of the disintegration/ dissolution of the family economic unit. Drawing again from the Foucaldian feminist critique, potential shame and impacts on children, that is, women's caring, may be a source of disempowerment for some women, who have feelings of disquiet within their family economic units, but remain in these situations despite being aware that there are alternatives.

Knowledge (power) claims

A critical analysis of industry language "us" and "them" reveals a discursive struggle between fisher(men) and management over attempts to claim True knowledge. Fisheries management claims privileged (masculine) scientific expert knowledge/ power legitimated by education, legislation and organisational authority. In contrast, fisher(men) claim "local" knowledge/ power legitimated through their (masculine privileged) occupational status and experience. While "I experience therefore I know" is a credible and substantial basis for a claim to knowledge (Harding 2004), the increasing power of fisheries management agencies to permit fishers to continue fishing or force them to cease fishing, means that fisher(men) are marginalised in this conflict.

The alleged embodied nature of the knowledge of fisher(men) is associated with the act of harvesting. However, this knowing may be attributed to men's enhanced opportunities for gaining the harvesting experience and skills through patrilineal inheritance and masculine occupational privileging. This knowledge/ power claim reflects and perpetuates masculine occupational privileging and women's exclusion from harvesting. As women are excluded from the act of harvesting, they are simultaneously excluded from the opportunity to "know". Women's lack of knowing was one explanation provided by participants as to why women do not fish. More significantly, this knowledge/ power claim provides (fisher)men with the *credibility* to claim stakeholder status. Many participants, including several male industry organisation and fisheries management representatives, feel that women do not "know" enough about fishing to participate in industry decision making.

I argue that as a result of the gendered constructions of "fisherman" and "fisherman's wife" and the associated discursive division of labour, women in fishing are more likely than men in fishing to engage in activities which maintain businesses, families and communities. This results in many women having a privileged (subjugated) knowledge that correlates to the social issues in sustainability. However, the activities women engage in are constructed as "natural" for wives/ mothers/ women. Unlike fisher(men), who are privileged with occupational status, and fisheries management, who are educationally qualified and implement decisions grounded in legislation, women have no *legitimate* basis to claim knowledge/ power. I argue that the possibility for women's substantial (subjugated) knowledge to be revealed and to form the basis for a standpoint for women in fishing is becoming increasingly unlikely in the current industry climate.

An "economically efficient" "sustainability" "crisis"

Since the early 1990s, the industry has been characterised by increasing regulation, which in turn has necessitated increased fisheries management effort. Current industry changes are being driven by the quest for sustainable industries underpinned by bio economic fisheries management and the principles of Ecologically Sustainable Development (ESD) in the broader climate of a neo liberal state. The language of "crisis" is prevalent. This language implies the need for immediate action by someone who "Knows" what they are doing and what to do. Furthermore, responsibility for the "crisis" is positioned with fisher(men)-the crisis is a result of fisher(men) over fishing for individual economic gain. Fishermen are deemed not competent to manage, or contribute to managing, the "crisis" and need to be managed. The knowledge/ power of management backed by legislation prevails over the local knowledge of "fishermen".

The language prominent in fisheries management discourse – for example, "economic sustainable development", "environmental management systems", "pressure", "competition", "cost recovery", "professionalism" and "economic efficiency"-reflects the exercising of power by fisheries management to construct the Truth that these terms are fundamentally reliant and interdependent on each other to resolve the "crisis". In addition, there is an illusive and constant use of the terms "more" and "increasing". Fisher(men) are instructed there is a need for them to become *more* "economically efficient" and "professional". Not only are the intended meanings of these terms not explained by management, but there is no quantification on how much more economically efficient and professional fishers need to become. How much "more" is enough? And what are the consequences for those who do not become professional enough?

I argue that deconstructing prevalent language and exploring meanings and practices in fisheries management, exposes the economic rationalist political nature of a seemingly apolitical fisheries management discourse. For example, the emerging subject position of "professional fisherman" implies a formalising of occupational status. This term can also be read as "exclusionary" where those who are not "professional" are not deemed competent to be in the industry. Attaining the status of "professional fishermen" for some fisher(men) who do not fit the construction of an economically efficient, educated, technically adept worker is difficult for those in smaller family fishing businesses, who may lack the education, personal resources and literacy levels (particularly European fishing families) to interpret and implement increasing amounts of complex administrative requirements. From management's point of view, "unprofessional" "uneconomically efficient" lifestyle family fishing businesses are breaching the emerging dominant discourse, which constructs the "right kind of" fishing businesses as professional and economically efficient.

My contention that there is a drive by fisheries management to displace smaller family fishing businesses is also substantiated by the way fisheries management *conveys* meaning. The majority of women in this research discuss the increasing amount of written text they receive which outlines regulation changes. There are also substantial penalties for not complying with these changes. Several women spoke of instances where the increasing amount and complexity of text meant that fisher(men) were not able to grasp the intended meaning, unintentionally did not comply and were heavily penalised. The "crisis"–that is, the need to act immediately-may be used to justify the expediency of fisheries management decision making, advice of regulation changes to fishing businesses, and enforcement. In contrast to larger corporate fishing businesses, which are able to employ people in specialised roles to interpret regulations and ensure they are complied with, smaller family fishing businesses are marginalised. Furthermore, family fishing businesses with no experience of negotiating or interpreting text in the way it is intended by fisheries management are deemed as inadequate and as *not competent to be involved in the fisheries*. Fisher(men) are blamed and the problem is seen as an individual and not a structural problem.

This argument also lends itself to an alternate reading of management's claim that the fishing industry is "apathetic". Indeed, it may be that management's poor consultation processes and their failure to incorporate the experiential knowledge of industry into decision making processes means men and women in fishing are disempowered and unable to "keep up" with management. In other words, industry is constructed as apathetic as they do not participate in the way which management deems fit. Again, instead of fisheries management reviewing their consultation processes, which would indicate a genuine commitment to incorporating the knowledge of men and women in fishing, responsibility for the lack of participation is positioned with men and women in fishing. This suggests that management's decision making processes are inadequate. This is concerning as decision making processes are essential to ESD (Fletcher 2002).

This also raises the issue of how well fisheries management incorporates other key components in their application of ESD, which is the formally adopted sustainability discourse in natural management resource in Australia. Two further issues of concern are, firstly, fisheries managements' failure to adequately consider social issues in decision making, and secondly, their inability to integrate environmental, economic and social issues. While ESD espouses the importance of the "social" in sustainable, its lack of attention reflects its discursive positioning as a secondary consideration after economic and environmental factors. In many cases, fisheries management exercise their knowledge/ power to justify this prioritising through the language of "crisis" where, in order to manage crisis, environmental and economic require immediate action. I argue that fisheries management is failing to adequately apply their definition of sustainability. In the majority of instances, there is only rhetoric in terms of the value placed on the social component of ESD. Furthermore, the prioritising of the environmental and economic components of ESD, and management's failure to apply an *integrated* approach to sustainability which does not "carve up" sustainability into three separate components, also reflects the Western, masculine, rationalist, scientific reductionist discourse that characterises fisheries management (Holm 1996).

This is in contrast to the way many women in this research discuss sustainability where language such as "holistic" and "social impacts" are prevalent. Even where women do not use this language, through the telling of stories about increasing workloads, stress, suicides and families exiting the industry and relocating from small towns as a result of management decisions, my interpretation is that these women conceptualise sustainability as essentially involving social issues. The majority of women in this research prioritise social issues equally with environmental and economic issues and re/construct "the industry" as more than "fishermen" harvesting, and instead as involving women, men, families and communities.

Implications for women in the current climate

There are significant implications for women in fishing in the current industry climate. Firstly, as women lack stakeholder status, they are not involved in decision making. The subject position of "stakeholder" in the commercial fishing industry aligns with the subject position of "fishermen". Both industry organisations and fisheries management fails to acknowledge women's activities as industry contributions. In industry, men's occupational privileging confers stakeholder status to (fisher)men. Revealing a similar masculine dominant management discourse, this construction is also reflected and perpetuated in fisheries management, where the terms "fisherman" and "stakeholder" are interchanged. While broader discourses of political correctness and equity issues in bureaucracies have influenced text and there is an increasing use of the term "fisher" in fisheries management and industry text, often the context implies a masculine meaning. This is evident when "fisher" is used in conjunction with masculine pronouns or refers exclusively to the practice of harvesting. Furthermore, there are still significant textual "slip ups" whereby stakeholders are directly named as "fishermen". In a recent social assessment study in the rock lobster fishery, women in fishing were referred to as "fishermen's housewives" (Huddleston and Drew 2003). Management's economically driven approach to "sustainable" management in the context of "crisis" does not genuinely prioritise issues of inclusive consultation and participatory decision making.

In the context where the knowledge/ power/ truth of fisheries management surpasses the knowledge/ power/ truth of fisher(men), women in fishing are uniting with their partners against management to ensure the survival of their family economic units. This perpetuates the secondary positioning of women in fishing and reduces the opportunities for them to be granted stakeholder status as when they demonstrate activism, it is constructed as "us" (the family economic unit) versus "them" (fisheries management). Women are unlikely to claim discursive space to redress their lack of stakeholder status, as doing so may potentially detract from a unified industry position and threaten their family economic unit.

Secondly, as the industry is constructed of "fishermen" who harvest, women, families and communities are not adequately considered in terms of the social impacts of fisheries management decisions. Indeed, the social impacts of management decisions are problematised at the local family fishing business level instead of at a structural level. For example, given fisheries management's failure to apply an integrated concept of sustainability, management may claim that certain decisions must be made based primarily on environmental and economic grounds due to the "crisis". The negative social impacts of decisions are viewed as an unfortunate consequence of the urgent need to restructure, which is necessary due to the crisis *caused* by fishers. While fisheries management may provide some form of compensation, for example, structural adjustment packages, aside from this, individual fisher(men) and families need to manage the social impacts.

Thirdly, men's occupational privileging, and particularly in patrilineal inheritance practices, further marginalises women's ownership status in the privatising of resources. The industry is becoming characterised by the introduction of market driven resource allocation strategies such as Individual Transferable Quotas (ITQs). In many fisheries, licenses are either legally unable to be split, or become economically unviable if they are split. This is resulting in an increasingly closed industry, where entry is restricted to family assistance/ business transference or the possession of substantial capital. As fishing is reserved as an occupation for men, I argue that this perpetuates the practice of patrilineal inheritance and continues to marginalise women's ownership status and constrain their entry to the industry through marriage.

Fourthly, given the need for women to unite with their partners in order to ensure the survival of their family businesses in the current climate, the opportunities for women to both claim and be granted discursive space are reduced.

Fifthly, women in fishing are increasingly excluded from the opportunity to harvest fish as the subject position of "professional fishermen" emerges and as opportunities for unpaid harvesting work decrease as more corporate "professional" fishing businesses become more prevalent. Finally, and at the most fundamental level, as fisheries management and regulation is ensuring that the industry is comprised of those who are "economically efficient", the wealth amongst fishing businesses is increasing. Women in family fishing businesses which are financially insecure are working harder in an endeavour to ensure their businesses and families survive, and an increasing number of women in fishing are exiting the industry as the number of smaller family fishing businesses decline. Given that many of these women and men in fishing who exit the industry may be involved in 2nd, 3rd and even 4th generation fishing businesses, may be ethnic, and may also live in rural areas, the potential social impacts for women and these families are extensive.

I argue that industry women in fishing are likely to become increasingly marginalised in the current management climate in the quest for more economically efficient (sustainable?) fisheries. However, women still demonstrate resistance and there is significant potential should women's subjugated knowledge be acknowledged, for women to enhance the future of their industry.

Resistance (power)

Some women in fishing demonstrate resistance against their discursive secondary subject positioning through a variety of mechanisms including: asserting their claim to be acknowledged as equal to "fishermen"; resisting undertaking the tasks associated with the subject position of "fishermen's wife"; claiming stakeholder status; becoming involved in industry decision making; and through collective action.

However, again there is an inherent complexity in the manner in which language and meaning intersect with differing women's subjectivities in exercising resistance. The complex nature of gendered industry relations are such that women in fishing may demonstrate collective resistance against fisheries management independently of their partners, while still retaining a secondary industry positioning. These women do not breach the dominant masculine industry discourse as they demonstrate resistance while simultaneously maintaining a secondary positioning of "supporting" their partners and their industry. However, these women *still* demonstrate resistance and perhaps this experience for some women may result in a level of empowerment. This complexity of positioning is also evident when considering WINSC. There can be no doubt that the vision, aims and practices of this organisation incorporate increasing the visibility and profile of women in fishing, including women's representation in industry decision making. However, vision, aims and practices of this organisation also incorporate activism to benefit the industry as a whole which, ironically, may perpetuate the primary industry positioning of "fishermen" in the industry.

Despite this, I argue that as a result of engaging in their daily "lifestyle" activities associated with the subject position of "fisherman's wife", many women in fishing have substantial knowledge of issues which are relevant to the social component of sustainability. The consequences of these activities being constructed as "normal" and "natural" for "fishermen's wives" ("wife"/ "woman") are that women's work is not recognised, and secondly, that women have no legitimate basis to claim knowledge. However, I argue that despite their lack of visibility, women in family businesses *are* industry stakeholders, and that the experience of women undertaking their daily activities provides them with a (subjugated) knowledge. This subjugated knowledge of women in fishing is uniquely placed to contribute to the knowledge gap on social issues in sustainability which is recognised as existing by some fisheries management agencies.

However, I also contend that it is becoming increasingly less likely that women's knowledge will be recognised. In order for women's knowledge to have some legitimate basis, women's activities would have to be acknowledged as vital "work" which contributes to the industry and women in fishing would have to be acknowledged as stakeholders. This would involve a reconstruction of "the industry" as involving more than both "fishermen", and more than the act of harvesting. This is not likely in an industry where both the fisheries management and the broader industry discourse are pervasively masculine, as acknowledging women's subjugated knowledge threatens this masculine power. Furthermore, the likelihood that women's knowledge will be recognised becomes less likely still where women unite with their partners while maintaining a secondary industry positioning to resist against fisheries management in order to attempt to ensure the survival of their family economic units.

While both women and men make vital contributions to their industry, businesses, families and communities, women in fishing lack visibility and are marginalised. I conclude that the Australian commercial fishing industry is locally, politically, systemically and structurally gendered. Furthermore, in the quest for "sustainable" fisheries, it is likely that women in family fishing businesses will be become increasingly marginalised. As management is charged with implementing legislation in accordance with the principles of ESD, it is clearly *the responsibility of management to redress* both the increasing marginalisation of smaller family businesses, and women within these businesses. Perhaps implementing strategies in accordance with fisheries management's own definition of ESD to address these inequities would ensure a more "sustainable" industry.

Further Research

As a result of this research, several avenues for further research emerge. Firstly, given the association between "sustainability", "economic efficiency" and the increasing corporatisation of fisheries, research that deconstructs the hierarchical relationship between man/ nature (Plumwood 1993) and reevaluates the purpose and practice of fishing could be beneficial to inform future work on sustainability in fisheries. Secondly, this research exposes a political agenda inherent in the discourse of sustainability. It would be enlightening to conduct an action research project in a fishery where management was prepared or attempting to apply a more integrated approach to implementing sustainable management, which incorporated the knowledge of all stakeholders including fisheries management, and women and men in fishing. Thirdly, while this research is a "snapshot" at one point in time, it would be politically powerful to undertake a longitudinal study of the social impacts where fisheries have undergone adjustments, restructures and closures. This is particularly pertinent as participants involved in this research had not yet exited the industry, yet discussed the trauma they had seen for those who had.

Fourthly, based on social justice principles, further research is needed to explore the absence of Aboriginal people from commercial fishing in Australia. Fifthly, research which explores in more detail the avenues for resistance for smaller family fishing businesses in the current climate, and which also exposes the political agendas beyond the language of "sustainability", could provide assistance to a range of primary industries, including agriculture, in specific social and historical contexts. This could be achieved by research which explores both gender and the "social" in sustainable in legal challenges by fishing businesses against fisheries management agencies which would provide significant insight on how these issues are prioritised in legal discourses, and inform powerful issues strategies for future action. Finally, the need for a feminist exploration of socio-economic impacts and relating to climate change as the latest Australian commercial fishing industry 'crisis' is critical.

A Final Word

This research provides the tools to critically view the marginalisation of women in fishing, the evidence required for political action, and the basis for women in fishing to claim a standpoint. It also signifies the start of a process where women in fishing are increasingly likely to be supported in achieving a standpoint, and as such, puts fisheries management agencies and industry organisations on notice that gender inequity needs to be addressed in the Australian commercial fishing industry.

BIBLIOGRAPHY

Acker, Joan, Kate Barry and Joke Esseveld. 1996. "Objectivity and Truth: Problems in Doing Feminist Research." In *Feminism and Social Change: Bridging Theory and Practice*, edited by Heidi Gottfried, 61-68. Urbana: University of Illinois Press.

Agarwal, Bina. 1992. "The Gender and Environment Debate." *Feminist Studies*. 18 (a):119-158.

Agriculture, Fisheries and Forestry Australia (AFFA). 1999. *Guidelines for Reaching Our Clients: Women*. Canberra, ACT: AFFA.

—. 2003. *Looking to the Future: A Review of Commonwealth Fisheries Policy*. Canberra, ACT: AFFA.

Allison, Charlene J. 1988. "Women Fishermen in the Pacific Northwest." In *To Work and to Weep: Women in Fishing Economies*, edited by Jane Nadel-Klein and Donna Lee Davis, 230-260. Newfoundland: Memorial University of Newfoundland.

Allison, Charlene J., Sue-Ellen Jacobs and Mary A. Porter. 1989. *Winds of Change: Women in Northwest Commercial Fishing*. Seattle and London: University of Washington Press.

Alston, Margaret. 1990. "Farm Women and their Work." In *Rural Women Keypapers Number 1*, edited by Margaret Alston, 29-41. Wagga Wagga: Centre for Rural Social Research.

—. 1991. "Women's Place in the Future of Agriculture." In *Family Farming Keypapers Number 2*, edited by Margaret Alston, 97-105. Wagga Wagga: Centre for Rural Social Research.

—. 1994a. "Feminism and Farm Women." In *Country Women at the Crossroads: Perspectives on the Lives of Rural Australian Women in the 1990s,* edited by Margaret-Ann Franklin, Leonie M. Short and Elizabeth K. Teather, 24-31. Armidale: University of New England Press.

—. 1994b. "Women and the Rural Crisis." In *Country Women at the Crossroads: Perspectives on the Lives of Rural Australian Women in the 1990s,* edited by Margaret-Ann Franklin, Leonie M. Short and Elizabeth K. Teather, 11-18. Armidale: University of New England Press.

—. 1995a. "Women and their Work on Australian Farms." *Rural Sociology* 60 (3):521-532.

—. 1995b. *Women on the Land: The Hidden Heart of Rural Australia.* Australia: UNSW Press.

—. editor. 1998a. *Australian Rural Women Towards 2000.* Wagga Wagga: Centre for Rural Social Research.

—. 1998b. "Farm Women and their Work: Why is it Not Recognised?" *Journal of Sociology* 34 (1):23-34.

—. 1998c. *Leading Women in Agriculture.* Wagga Wagga: Centre for Rural Social Research.

—. 1998d. "The "Othering" of Rural Women." In *Australian Rural Women Towards 2000,* edited by Margaret Alston, 31-37. Wagga Wagga: Centre for Rural Social Research.

—. 1998e. "There are Just no Women Out There: How the Industry Justifies the Exclusion of Women from Agricultural Leadership." *Rural Society* 8 (3):197-209.

—. 2000. *Breaking Through the Grass Ceiling: Women, Power and Leadership in Agricultural Organizations.* Australia: Harwood Academic Publishers.

—. 2003. "Australian Rural Women: Mainstreaming the Agenda." In *Setting the Agenda for Rural Women: Research Directions, Conference Proceedings and Recommendations*, edited by Rowan O'Hagan, Margaret Alston and Shelley Spriggs. Wagga Wagga: Centre for Rural Social Research.

—. 2006. "'I'd Just Like to Walk Out of Here': Australian Women's Experiences of Drought." *Sociologica Ruralis* 46 (2):154-170.

Alston, Margaret and Jenny Kent. 2004. *Social Impacts of Drought: A Report to NSW Agriculture.* Wagga Wagga: Centre for Rural Social Research.

Alston, Margaret and Jenny Kent. 2006. *Impacts of Drought on Educational Access for Rural and Remote Young People: A Report to DEST.* Wagga Wagga: Centre for Rural Social Research.

Alston, Margaret and Jane Wilkinson. 1998. "Australian Farm Women-Shut Out or Fenced In? The Lack of Women in Agricultural Leadership." *Sociologica Ruralis* 38 (3):391-408.

Ardovini-Brooker, Joanne. 2001. "The Debates and Unresolved Issues Surrounding Feminist Research and Its Distinction from Mainstream Research." *Advancing Women in Leadership Journal* Winter. Accessed March 20 2002. www.advancingwomen.com/awl/winter2001/Ardovini-Brooker.htm.

Armstead, Cathleen. 1995. "Writing Contradictions: Feminist Research and Feminist Writing." *Women's Studies International Forum* 18 (5/6):627-636.

Armstrong, Karen. 1978. "Rural Scottish Women: Politics without Power." *Ethnos* 43 (1-2):51-72.

Aslin, Heather J., Robin D. Connor and Melanie Fisher. 2001. *Sharing in the Catch or Cashing in the Share? Social Impacts of Individual Transferable Quotas and the South East Fishery*. Canberra: Bureau of Rural Sciences.

Aslin, Heather J., Trevor Webb and Melanie Fisher. 2000. *Fishing for Women: Understanding Women's Roles in the Fishing Industry*. Canberra: Bureau of Rural Science.

Australian Bureau of Agricultural and Resource Economics (ABARE). 1998. *People in Farming*. Canberra, ACT: ABARE.

Australian Fisheries Management Authority (AFMA). 2005. *AFMA Update* 2 (24).

Bailey, ME. 1993. "Foucauldian Feminism: Contesting Bodies, Sexuality and Identity." In *Up Against Foucault: Explorations of Some Tensions between Foucault and Feminism*, edited by Caroline Ramazanoglu, 99-122. London: Routledge.

Baker, Rachel and Rachel Hinton. 1999. "Do Focus Groups Facilitate Meaningful Participation in Social Research?" In *Developing Focus Group Research: Politics, Theory and Practice*, edited by Rosaline S. Barbour and Jenny Kitzinger, 79-99. London: SAGE.

Beasley, Chris. 1999. *What is Feminism, Anyway? Understanding Contemporary Feminism*. Australia: Allen and Unwin.

Beder, Sharon. 2000. "Costing the Earth: Equity, Sustainable Development and Environmental Economics." *New Zealand Journal of Environmental Law* 4:227-243.

Beilin, Ruth. 1998. "Inside Out: Exploring the Connection between Women's Life Stories and Landscape." *Rural Society* 8 (3):165-183.

Bell, JH. and US Pandey. 1989. "Gender-Role Stereotypes in Australian Farm Advertising." *Media Information Australia* 51:45-49.

Bhatta, Gambhir. 2001. "Of Geese and Ganders: Mainstreaming Gender in the Context of Sustainable Human Development." *Journal of Gender Studies* 10 (1):17-32.

Bhavnani, Kum-Kum. 1993. "Tracing the Contours: Feminist Research and Feminist Objectivity." *Women's Studies International Forum* 16 (2):95-104.

Binkley, Marian. 1996. "Nova Scotia Fishing Families Coping with the Fisheries Crisis." *Anthropologica* 38:197-219.

—. 2000. "'Getting By' in Tough Times: Coping with the Fisheries Crisis." *Women's Studies International Forum* 23 (3):323-332.

—. 2005. "The Bitter End: Women's Crucial Role in the Nova Scotia

Coastal Fishery." In *Changing Tides: Gender, Fisheries and Globalization,* edited by Barbara Neis, Marian Binkley, Siri Gerrard and Maria C. Maneschy. Halifax: Fernwood Publishing.

Board, Helen. 1998. "Rural Women: Policy and the Process of Change in Australian Agriculture." In *Australian Rural Women Towards 2000*, edited by Margaret Alston, 37-47. Wagga Wagga: Centre for Rural Social Research.

Bourdieu, Pierre. 2001. *Masculine Domination*. Cambridge: Polity Press.

Bourke, Lisa and Stewart Lockie. 2001. "Rural Australia: An Introduction." In *Rurality Bites: The Social and Environmental Transformation of Rural Australia*, edited by Lisa Bourke and Stewart Lockie, 1-15. Australia: Pluto Press.

Bowling, Jill and Brian Martin. 1985. "Science: A Masculine Disorder?" *Science and Public Policy* 12 (6):308-316.

Braidotti, Rosi, Ewa Charkiewiwcz, Sabine Hausler and Saskia Wieringa, editors. 1994. *Women, the Environment and Sustainable Development: Towards a Theoretical Synthesis*. London: Zed Books in conjunction with INSTRAW.

Bradshaw, Matt. 2004. "The Market, Marx and Sustainability in a Fishery." *Antipode* 36 (1):66-85.

Brandth, Berit, and Marit S. Haugen. 1998. "Breaking into a Masculine Discourse: Women and Farm Forestry." *Sociologia Ruralis* 38 (3):427-442.

Brandth, Berit, and Marit S. Haugen. 2000. "From Lumber Jack to Business Manager." *Journal of Rural Studies* 16:343-355.

Brandth, Berit, and Marit S. Haugen. 2005. "Doing Rural Masculinity: From Logging to Outfield Tourism." *Journal of Gender Studies* 14 (1):13-22.

Broad, Lucy. 1998. "The Untapped Resource." In *Australian Rural Women Towards 2000*, edited by Margaret Alston, 61-67. Wagga Wagga: Centre for Rural Social Research.

Bryant, Lia. 1991. "Family Farm Displacement." In *Family Farming Keypapers Number 2,* edited by Margaret Alston, 81-97. Wagga Wagga: Centre for Rural Social Research.

—. 1999. "The Traditionalization of Occupational Identities in Farming in South Australia." *Sociologica Ruralis* 39 (2):237-261.

—. 2001. "A Job of One's Own." In *Rurality Bites: The Social and Environmental Transformation of Rural Australia*, edited by Lisa Bourke and Stewart Lockie, 214-227. Australia: Pluto Press.

Burch, David, Geoffrey Lawrence, Roy Rickson and Jasper Goss. 1998. *Australasian Food and Farming in a Globalised Economy: Recent*

Developments and Future Prospects. Melbourne: Department of Geography and Environmental Science, Monash University.

Cain, Maureen. 1993. "Foucault, Feminism and Feeling: What Foucault Can and Cannot Contribute to Feminist Epistemology." In *Up Against Foucault: Explorations of Some Tensions between Foucault and Feminism*, edited by Caroline Ramazanoglu, 73-98. London: Routledge.

Campbell, Rebecca and Sharon Wasco. 2000. "Feminist Approaches to Social Science: Epistemological and Methodological Tenets." *American Journal of Community Psychology* 28 (6):773-792.

Caton, A. and K. McLoughlin. 2000. *Fishery Status Reports 1999: Resource Assessments of Australian Commonwealth Fisheries.* Canberra, ACT: Bureau of Rural Sciences.

Chao, NH., CF. Chang and EY. Chang. 2002. "Responding to Globilization Trends in Fisheries: Assets of Professional Taiwanese Women. In *Global Symposium on Women in Fisheries: Sixth Asian Fisheries Forum 29 November 2001,* edited by Meryl J Williams, NH Chao, Poh Sze Choo, K Matics, MC Nandeesha, M Shariff, I. Siason, E Tech and JMC Wong, 65-77. Malaysia: ICLARM-The World Fish Centre.

Chiu, Lai-Fong and Deborah Knight. 1999. "How Useful are Focus Groups for Obtaining the Views of Minority Groups?" In *Developing Focus Group Research: Politics, Theory and Practice*, edited by Rosaline S. Barbour and Jenny Kitzinger, 99-113. London: SAGE.

Claridge, Ciel. 1998. "Rural Women: Decision Making and Leadership within Environmental and Landcare Groups." *Rural Society* 8 (3):183-197.

Collins, Patricia Hill. 1991. "Learning From the Outsider within: The Sociological Significance of Black Feminist Thought." In *Beyond Methodology: Feminist Scholarship as Lived Research*, edited by Mary M. Fonow and Judith A. Cook, 35-59. Bloomington: Indiana University Press.

—. 2004a. "Learning From the Outsider within: The Sociological Significance of Black Feminism." In *The Feminist Standpoint Theory Reader: Intellectual and Political Controversies*, edited by Sandra Harding, 103-126. New York: Routledge.

—. 2004b. "Comment on Heckman's 'Truth and Method: Feminist Standpoint Theory Revisited': Where's the Power?" In *The Feminist Standpoint Theory Reader: Intellectual and Political Controversies*, edited by Sandra Harding, 247-254. New York: Routledge.

Commonwealth of Australia (CoA). 1991. *ESD Working Groups Final Report-Fisheries.* Canberra, ACT: Australian Government Publishing

Service.

Connell, Robert W. 1995. *Masculinities*. Cambridge: Polity Press.

—. 2003. "Masculinities, Change, and Conflict in Global Society: Thinking about the Future of Men's Studies." *Journal of Men's Studies* 11 (3):249-266.

Cook, Judith A. and Mary M. Fonow. 1990. "Knowledge and Women's Interests: Issues of Epistemology and Methodology in Feminist Sociological Research." In *Feminist Research Methods*, edited by Joyce M. Nielsen, 69-94. Boulder: Westview Press.

Crean, Kevin and David Symes. 1996. "Conclusion." In *Fisheries Management in Crisis*, edited by Kevin Crean and David Symes, 197-206. Oxford: Fishing News Books.

Crist, Eileen. 2007. "Beyond the Climate Crisis: A Critique of Climate Change Discourse." *Telos* 141:29-55

Dallimore, Elise. 2000. "A Feminist Response to Issues of Validity in Research." *Women's Studies in Communication* 23 (2):157-181.

Davidson, AP. 2001. "Farming Women and the Masculinisation of Farming Practices." In *Rurality Bites: The Social and Environmental Transformation of Rural Australia*, edited by Lisa. Bourke and Lockie Stewart, 204-214. Australia: Pluto Press.

Davidson, Penny, and Rosemary Black. 2001. "Women and Natural Resource Management: Finding a More Balanced Perspective." *Society and Natural Resources* 14 (8):645-656.

Davis, Anthony, and Conner Bailey. 1996. "Common in Custom, Uncommon in Advantage: Common Property, Local Elites and Alternative Approaches to Fisheries Management." *Society and Natural Resources* 9:251-265.

Davis, Dona Lee. 1983. "Women the Worrier: Confronting Archetypes of Stress." *Women's Studies* 10 (2):135-146.

—. 1993. "When Men Become Women: Gender Antagonism and the Changing Sexual Geography of Work in Newfoundland." *Sex Roles* 29 (7-8):1-18.

Davis, Dona Lee, and Siri Gerrard. 2000. "Introduction-Gender and Resource Crisis in the North Atlantic Fisheries." *Women's Studies International Forum* 23 (3):279-286.

Davis, Dona Lee. 1986. "Occupational Community and Fishermen's Wives in a Newfoundland Fishing Village." *Anthropological Quarterly* 59 (3):129-142.

—. 2000. "Gendered Cultures of Conflict and Discontent: Living 'the Crisis' in a Newfoundland Community." *Women's Studies International Forum* 23:343-353.

Davis, Dona Lee, and Jane Nadel-Klein. 1992. "Gender, Culture and the Sea: Contemporary Theoretical Approaches." *Society and Natural Resources* 5:135-147.

de Monchaux, Annie. 1999. *Cray Tales*. Fremantle, WA: Nice to be Nice.

Dempsey, Ken. 1992. *A Man's Town: Inequality between Men and Women in Rural Australia.* Melbourne: Oxford University Press.

—. 1994. "The Oppression Of Rural Women in Australia." In *Country Women at the Crossroads: Perspectives on the Lives of Rural Australian Women in the 1990s,* edited by Margaret-Ann Franklin, Leonie M. Short and Elizabeth K. Teather, 41-51. Armidale: University of New England Press.

Department Agriculture, Fisheries and Forestry (DAFF). 2008. *Employment in commercial fishing, seafood processing and seafood wholesaling*. Accessed February 20 2011. http://www.daff.gov.au/brs/fisheries-marine/info/fishery_status_reports_2008/chapters/overview/economic_status/employment_in_commercial_fishing,_seafood_processing_and_seafood_wholesaling

Department of Natural Resources and Environment (DNRE). n. d. *Needs Assessment and Analysis of Women in the Seafood Industry.* Victoria: Department of Natural Resources and Environment.

Deveaux, Monique. 1994. "Feminism and Empowerment: A Critical Reading of Foucault." *Feminist Studies* 20(2):23-247.

Dixon, Richard, Roger Lowery, James Sabella and Marcus Hepburn. 1984. "Fishermen's Wives: A Case Study of a Middle Atlantic Fishing Community." *Sex Roles*, 10 (1-2): 33-52.

Dowling, Jane. 2006a. "Fishing for Power? Women in the Fishing Industry in Australia." In *Fluid Bonds: Views on Gender and Water*, edited by Kuntala Lahiri-Dutt, 138-150. Kolkata: Stree.

—. 2006b. *Teleworking: Building Capacity, Sustaining Rural and Regional Communities and Creating Opportunities for Women.* Australia: Foundation for Australian Agricultural Women.

Ecologically Sustainable Development Steering Committee. 1992. *National Strategy for Ecologically Sustainable Development*. Accessed April 22 2005. www.environment.gov.au/esd/national/nsesd/strategy/index.html-12k.

England, Kim V. 1994. "Getting Personal: Reflexivity, Positionality, and Feminist Research." *Professional Geographer* 46 (1):80-89.

European Commission. 2000. *Sustainable Governance: Institutional and Procedural Aspects of Sustainability*. Luxemburg: Office for Official Publications for the European Communities.

Fawcett, Barbara, Brid Featherstone, Jan Fook and Amy Rossiter, editors. 2000. *Practice and Research in Social Work: Postmodern Feminist Perspectives.* London: Routledge.

Federal Ministry for the Environment, Nature Conservation and Nuclear Safety (BMU), Heinrich Boell Foundation (HBF) and UNED Forum. 2001. *Gender perspectives for Earth Summit 2002 – energy, transport, information for decision making, Report on the International Conference at Jagdschloss Glieniche, Germany, 10-12 January 2001.* Germany: HBF & BMU. Accessed March 2 2006. www.worldsummit2002.org/publications/genderpersp2002.pdf

Ferrari, Christine and Jennifer McKinnon, editors. 1998. "A Selection of Abridged Papers: Second International Conference on Women in Agriculture." *Rural Society* 8 (3):271-286.

Fink, Deborah. 1990. "Farm Wives and Agrarianism in the United States." In *Rural Women Keypapers Number 1,* edited by Margaret Alston, 1-11. Wagga Wagga: Centre for Rural Social Research.

Fisheries Research and Development Corporation (FRDC). 2000. *Investing for Tomorrow's Fish: The FRDC'S Research and Development Plan, 2000-2005.* Canberra, ACT: FRDC.

—. 2001. *From Antarctica to the Tropics: A Snapshot of the Australian Fishing Industry.* Canberra, ACT: FRDC.

Fishing Crisis. 2006. Queensland: ABC Stateline, April 21. Television broadcast.

Fletcher, WJ. 2002. *Policy for the Implementation of Ecologically Sustainable Development for Fisheries and Aquaculture in Western Australia.* Western Australia: Department of Fisheries.

Fletcher, WJ, Chubb, C, McCrea, J, Caputi, N, Webster, F, Gould, R and Bray, T. 2005. *Western Rock Lobster Fishery, ESD Report Series No. 4.* Western Australia: Department of Fisheries.

Fletcher, WJ. 2006. "Frameworks for Managing Marine Resources Using Ecosystem Approaches: How do they Fit Together and Can they be Useful?" *Bulletin of Marine Science* (under review). Accessed February 4 2007. www.fisheries-esd.com/c/pubs/index.cfm-31k

Fletcher, WJ, Chesson, J, Fisher, M, Sainsbury, KJ, Hundloe, T, Smith, ADM and Whitworth, B. 2002. *National ESD Reporting Framework for Australian Fisheries: The "How To" Guide for Wild Capture Fisheries, FRDC Project 2000/145.* Canberra.

Fonow, Mary M., and Judith A. Cook, editors. 1990. *Beyond Methodology: Feminist Scholarship as Lived Research.* Bloomington: Indiana University Press.

Foucault, Michel. 1980. *Power/Knowledge: Selected Writings and Other*

Writings 1972-1977. New York: Pantheon Books.

France, Victor, Larry Mitchell and Alison Wright. 1998. *Abrolhos Islands*. Western Australia: Fremantle Arts Centre Press.

Franklin, Margaret-Ann. 1989. "Aboriginal Women in Rural Australia." In *Women in Rural Australia,* edited by Kerry James, 116-133. Queensland: University of Queensland Press.

Garnaut, Jayne, Caroline Rasheed and Gil Rodriguez. 1999. *Farmers at Work: The Gender Division*. Canberra: ABARE.

GEC Diesels Australia Ltd. 1976. Brochure insert in *Australian Fisheries* 35(8).

Gendron, Louise, Rejeanne Camirand and Josée Archaneault. 2000. "Knowledge-Sharing Between Fishers and Scientists: Towards a Better Understanding of the Status of Lobster Stocks in the Magdalen Islands," In *Finding Our Sea Legs: Linking Fishery People and their Knowledge with Science and Management*, edited by Barbara Neis and Lawrence Felt, 56-71. Newfoundland: ISER.

Gerrard, Siri. 2000. "The Gender Dimension of Local Festivals: The Fishery Crisis and Women's and Men's Political Actions in North Norwegian Communities." *Women's Studies International Forum* 23 (3):299-309.

Gibbs, M. 1989. "Decision-Making on Pastoral Properties." In *Women in Rural Australia*, edited by Kerry James, 30-43. Queensland: University of Queensland Press.

Gibson, D, J Baxter and C Kingston. 1990. "Beyond the Dichotomy: The Paid and Unpaid Work of Rural Women." In *Rural Women Keypapers Number 1,* edited by Margaret Alston, 11-29. Wagga Wagga: Centre for Rural Social Research.

Gill, June. 1999. *Information Booklet on Meeting Roles and Procedures for Regional WIN.* (1999/356 Stage One A).

Gillham, Bill. 2000. *Case Study Research Methods*. London: Continuum.

Gorelick, Sherry. 1996. "Contradictions of Feminist Methodology." In *Feminism and Social Change: Bridging Theory and Practice*, edited by Heidi Gottfried, 23-45. Urbana: University of Illinois Press.

Gorton, Stan. 2001. "Roles for Women in Fishing". *Port Lincoln Times*, 24 May, 9.

Gottfried, Heidi. 1996. "Introduction-Engaging Women's Communities: Dilemmas and Contradictions in Feminist Research." In *Feminism and Social Change: Bridging Theory and Practice,* edited by Heidi Gottfried, 1-20. Urbana: University of Illinois Press.

—. editor. 1996. *Feminism and Social Change: Bridging Theory and Practice*. Urbana: University of Illinois Press.

Gough, Brendan, and Gareth Edwards. 1998. "The Beer Talking: Four Lads, a Carry Out and the Reproduction of Masculinities." *Sociological Review* 46:409-435.

Grace, Margaret, and June Lennie. 1998. "Constructing and Reconstructing Rural Women in Australia: The Politics of Change, Diversity and Identity." *Sociologica Ruralis* 38 (3):350-369.

Grafton, R. Quentin, Ragnar Arnason, Trond Bjorndal, David Campbell, Harry Campbell, Colin Clark, Robin Connor, Diane Dupont, Rognvaldur Hannesson, Ray Hilborn, James Kirkley, Tom Kompas, Daniel E. Lane, Gordon R. Munro, Sean Pascoe, Dale Squires, Stein Steinshamn, Bruce Turris and Quinn Weninger. 2005. *Incentive-Based Approaches to Sustainable Fisheries.* Canberra, ACT: Australian National University Economics and Environment Network.

Grafton, R Quentin, Tom Kompas and Kate Barkclay. 2006. *Fishing Futures, Policy Brief 1.* Canberra, ACT: Crawford School of Economics and Government, National University.

Gray, Howard. 1999. Book 2: *The Western Rock Lobster: A History of the Fishery, Panulirus Cygnus*. Geraldton: Westralian Books.

Gray, Ian. 1991. "Family Farming and Ideology: Some Preliminary Exploration." In *Family farming Keypapers Number 2*, edited by Margaret Alston, 55-69. Wagga Wagga: Centre for Rural Social Research.

Green, Judith and Laura Hart. 1999. "The Impact of Context on Data." In *Developing Focus Group Research: Politics, Theory and Practice*, edited by Rosaline S. Barbour and Jenny Kitzinger, 21-36. London: SAGE.

Greenbaum, Thomas L. 1998. *The Handbook for Focus Group Research.* 2nd ed. Thousand Oaks: SAGE.

Grimshaw, Jean. 1993. "Practices of Freedom." In *Up Against Foucault: Explorations of Some Tensions between Foucault and Feminism*, edited by Caroline Ramazanoglu, 51-72. London: Routledge.

Gronbech, Dagrunn. 2000. "Recycling the Past: Perspectives on Women, Households and Resource Management among Early 20th Century Fisher/Farmers in North Norway." *Women's Studies International Forum* 23 (3):355-361.

Hakim, Catherine. 1993. "Research Analysis of Administrative Records." In *Social Research: Philosophy, Politics and Practice,* edited by Martyn Hammersely, 131-146. London: Open University Press/ SAGE.

—. 2000. *Research Design: Successful Designs for Social and Economic Research.* 2nd ed. London: Routledge.

Hall-Arber, Madeleine. 1996. "Hear Me Speak: Italian and Portuguese Women Facing the Fishery Crisis." *Anthropologica* 38:221-248.

Hamel, James, Stephanie Dufour and Dominic Fortin. 1993. *Case Study Methods.* USA: SAGE.

Hannesson, Rognvaldur. 2004. *The Privatization of the Oceans.* Massachusetts: Massachusetts Institute of Technology.

Haraway, Donna. 1988. "Situated Knowledges: The Science Question in Feminism and the Privilege of Partial Perspective." *Feminist Studies* 14:575-599.

—. 2004. "Situated Knowledges: The Science Question in Feminism and the Privilege of Partial Perspective." In *The Feminist Standpoint Theory Reader: Intellectual and Political Controversies*, edited by Sandra Harding, 81-102. New York: Routledge.

Hardin, Garrett. 1968. "The Tragedy of the Commons." *Science* 162:1243-1248.

Harding, Sandra. 1987a. "Introduction-Is there a Feminist Method?" In *Feminism and Methodology*, edited by Sandra Harding, 1-15. USA: Indiana University and Open University Press.

—. 1987b. "Conclusion-Epistemological Questions." In *Feminism and Methodology,* edited by Sandra Harding, 181-191. USA: Indiana University and Open University Press.

—. 2004a. *The Feminist Standpoint Theory Reader: Intellectual and Political Controversies.* New York: Routledge.

—. 2004b. "Introduction: Standpoint Theory as a Site of Political, Philosophy and Science." In *The Feminist Standpoint Theory Reader: Intellectual and Political Controversies*, edited by Sandra Harding, 1-16. New York: Routledge.

—. 2004c. "Rethinking Standpoint Epistemology: What is "Strong Objectivity?" In *The Feminist Standpoint Theory Reader: Intellectual and Political Controversies*, edited by Sandra Harding, 127-140. New York: Routledge.

—. 2004d. "Comment on Heckman's "Truth and Method: Feminist Standpoint Theory Revisited": Standpoint Needs the Regimes of Truth and Reality." In *The Feminist Standpoint Theory Reader: Intellectual and Political Controversies,* edited by Sandra Harding, 255-262. New York: Routledge.

Hartsock, Nancy CM. 1987. "The Feminist Standpoint: Developing the Ground for a Specifically Feminist Materialism." In *Feminism and Methodology*, edited by Sandra Harding, 157-181. USA: Indiana University and Open University Press.

—. 2004a. "The Feminist Standpoint: Developing the Ground for a

Specifically Feminist Historical Materialism." In The *Feminist Standpoint Theory Reader: Intellectual and Political Controversies*, edited by Sandra Harding, 35-54. New York: Routledge.

—. 2004b. "Comment on Heckman's "Truth and Method: Feminist Standpoint Theory Revisited" In *The Feminist Standpoint Theory Reader: Intellectual and Political Controversies*, edited by Sandra Harding, 243-246. New York: Routledge.

Haslam-McKenzie, Fiona. 1998a. "Farm Women and the "F" Word", In *Australian Rural Women Towards 2000*, edited by Margaret Alston, 23-31. Wagga Wagga: Centre for Rural Social Research.

—. 2003a. "Strategies for Getting Women Back on Centre Stage." In *Setting the Agenda for Rural Women: Research Directions. Conference Proceedings and Recommendations*, edited by Rowan O'Hagan, Margaret Alston and Shelley Spriggs, 36-46. Wagga Wagga: Centre for Rural Social Research.

—. 2003b. "A Review of the Status of Women in Rural, Regional and Remote Australia: Identifying Leadership Strategy Successes and Challenges." In *Setting the Agenda for Rural Women: Research Directions. Conference Proceedings and Recommendations*, edited by Rowan O'Hagan, Margaret Alston and Shelley Spriggs, 133-150. Wagga Wagga: Centre for Rural Social Research.

Hassanein, Neva. 1997. "Networking Knowledge in the Sustainable Agriculture Movement: Some Implications of the Gender Dimension." *Society and Natural Resources* 10 (3):251-259.

Hekman, Susan J., 2004. "Truth and Method: Feminist Standpoint Theory Revisited." In *The Feminist Standpoint Theory Reader: Intellectual and Political Controversies*, edited by Sandra Harding, 225-242. New York: Routledge.

Herreria, E., A. Woodhead, R. Tottenham and C. Magpantay. 2004. *Social Profile of People Employed in Agriculture, Forestry and Fishing Industries*. Canberra: RIRDC.

Hesse-Biber, Sharlene, Christina Gilmartin, and Robin Lyndenberg, editors. 1999. *Feminist Approaches to Theory and Methodology: An Interdisciplinary Reader*. Oxford: Oxford University Press.

Hirschman, Nancy J. 2004. "Feminist Standpoint as Post Modern Strategy." In *The Feminist Standpoint Theory Reader: Intellectual and Political Controversies*, edited by Sandra Harding, 317-332. New York: Routledge.

Hobday, AJ., ES. Poloczanska, and RJ. Matear, editors. 2008. *Implications of Climate Change for Australian Fisheries and Aquaculture: a preliminary assessment. Report to the Department of Climate Change*. Canberra, Australia.

Hogan, Elisabeth. 1994. "Making Women Visible: Reflections on Working with Women in Agriculture in Victoria." In *Country Women at the Crossroads: Perspectives on the Lives of Rural Australian Women in the 1990s*, edited by Margaret-Ann Franklin, Leonie M. Short and Elizabeth K. Teather, 31-41. Armidale: University of New England Press.

Holm, Petter. 1996. "Fisheries Management and the Domestication of Nature." *Sociologica Ruralis* 36:177-188.

hooks, bell. 2004. "Choosing the Margin as a Space of Radical Openness." In *The Feminist Standpoint Theory Reader: Intellectual and Political Controversies*, edited Sandra Harding, 153-160. New York: Routledge.

Howell, LA. 2002. "Perspectives on Women in Fisheries in North America." In *Global Symposium on Women in Fisheries: Sixth Asian Fisheries Forum*, edited by MJ Williams, NH Chao, PS Choo, K Matics, MC Nandeesha, M Shariff, I Siason, E Tech and JMC Wong, 183-191. Malaysia: ICLARM-The World Fish Centre.

Huang, M. 2002. "HIV/AIDS Among Fishers: Vulnerability of their Partners." In *Global Symposium on Women in Fisheries: Sixth Asian Fisheries Forum*, edited by MJ Williams, NH Chao, PS Choo, K Matics, MC Nandeesha, M Shariff, I Siason, E Tech and JMC Wong, 49-55. Malaysia: ICLARM-The World Fish Centre.

Huddleston, Veronica, and Neil Drew. 2003. *The Assessment of Prospective Changes to Social Infrastructure as a Result of Rock Lobster Industry Management Initiatives*. Western Australia: Institute for Regional Development.

Hughes, Christina. 2002. *Key Concepts in Feminist Theory and Research*. London: SAGE.

Humm, Maggie. 1995. *The Dictionary of Feminist Theory*. Hertfordshire: Simon and Schuster.

Hutchings, Jeffrey A., and Mark Ferguson. 2000. "Links Between Fishers' Knowledge, Fisheries Science, and Resource Management: Newfoundland's Inshore Fishery the Northern Atlantic Cod, Gadus Morhua." In *Finding Our Sea Legs: Linking Fishery People and their Knowledge with Science and Management*, edited by Barbara Neis and Lawrence Felt, 82-110. Newfoundland: ISER.

International Labour Organisation (ILO). 1999. *Safety and Health in the Fishing Industry: Report for Discussion at the Tripartite Meeting on Safety and Health in the Fishing Industry*. Geneva: International Labour Office. Accessed June 8 2007.

http://www.ilo.org/public/english/dialogue/sector/techmeet/tmfi99/tmfir2.htm

Jagger, Alison M. 2004. "Feminist Politics and Epistemology: The Standpoint of Woman." In *The Feminist Standpoint Theory Reader: Intellectual and Political Controversies*, edited by Sandra Harding, 55-66. New York: Routledge.

James, Kerry. 1989. "Introduction." In *Women in Rural Australia*, edited by Kerry James, 1-20. Queensland: University of Queensland Press.

—. 1990. "Women's Decision-Making in Extended Family Farm Businesses." In *Rural Women Keypapers Number 1*, edited by Margaret Alston, 41-53. Wagga Wagga: Centre for Rural Social Research.

Jennings, Gayle and Daniela Stehlik. 2000. "Agricultural Women in Central Queensland and Changing Modes of Production." *Rural Society* 10 (1):63-79.

Jennings, Simon, Michel J. Kaiser and John D. Reynolds. 2001. *Marine Fisheries Ecology*. UK: Blackwell Science Ltd.

Jones, Andrew, Don Seville and Donella Meadows. 2002. "Resource Sustainability in Commodity Systems: The Sawmill Industry in the Northern Forest." *System Dynamics Review* 18 (2):171-204.

Jupp, Victor and Clive Norris. 1993. "Traditions in Documentary Research." In *Social Research: Philosophy, Politics and Practice*, edited by Martyn Hammersley, 37-52. London: Open University Press / SAGE.

Jusseit, Hans and E. Robinson. 2003. *Eastern Tuna and Billfish Fishery: Industry Code of Practice*. Queensland: East Coast Tuna Boat Owners.

Kaplan, Ilene M. 1988. "Women Who Go to Sea: Working in the Commercial Fishing Industry." *Journal of Contemporary Ethnography* 16 (4):491-314.

Kelly, Liz, Sheila Burton and Linda Regan. 1994. "Researching Women's Lives or Studying Women's Oppression? Reflections on What Constitutes Feminist Research." In *Researching Women's Lives from a Feminist Perspective,* edited by Mary Maynard and June Purvis, 27-48. London: Taylor and Francis.

Kitzinger, Jenny and Rosaline S. Barbour. 1999. "Introduction-The Challenge and Promise of Focus Groups." In *Developing Focus Group Research: Politics, Theory and Practice,* edited by Rosaline S. Barbour and Jenny Kitzinger, 1-21. London: SAGE.

Kubik, Wendee. 2005. "Farm Women: The Hidden Subsidy in Our Food." *Canadian Woman Studies* 24 (4):85-90.

Kura, Yumiko, Carmen Revenga, Eriko Hoshino and Greg Mock. 2004.

Fishing for Answers: Making Sense of the Global Fish Crisis. Washington: World Resources Institute.

Lambeth, Lyn, B. Hanchard, Heather Aslin, L. Fay-Sauni, P. Tuara, KD. Rochers and A. Vunisea. 2002. "An Overview of the Involvement of Women in Fisheries Activities in Oceania." In *Global Symposium on Women in Fisheries: Sixth Asian Fisheries Forum*, edited by MJ Williams, NH Chao, PS Choo, K Matics, MC Nandeesha, M Shariff, I Siason, E Tech and JMC. Wong, 127-145. Malaysia: ICLARM-The World Fish Centre.

Leach, P and L. Twycross. 2003. "A Family Tradition." In *Western Fisheries,* 34-36. Western Australia: Western Australian Department of Fisheries.

Lees, James. 1997. "The Future of the Family Farm." In *A Legacy under Threat? Family Farming in Australia*, edited by James Lees, 245-255. Australia: University of New England Press.

Liepins, Ruth. 1995. "Women in Agriculture: Advocates for a Gendered Sustainable Agriculture." *Australian Geographer* 26 (2):118-126.

—. 1996. "Reading Agricultural Power: Media as Sites and Processes in the Construction of Meaning." *New Zealand Geographer* 52 (2):3-10.

—. 1998. "Fields of Action: Australian Women's Agricultural Activism in the 1990s." *Rural Sociology* 63 (1):128-156.

—. 2001. "The Gendering of Farming and Agricultural Politics: A Matter of Discourse and Power." *Australian Geographer* 29 (3):371-388.

Little, Jo. 1994. "Gender Relations and the Rural Labour Process." In *Gender and Rurality*, edited by Sarah Whatmore, Terry Marsden and Phillip Lowe, 11-30. London: David Fulton Publishers.

Little, Jo and Carol Morris. 2005. *Critical Studies in Rural Gender Issues*, USA: Ashgate.

Lockie, Stewart. 2001b. "Agriculture and Environment." In *Rurality Bites: The Social and Environmental Transformation of Rural Australia,* edited by Lisa Bourke and Stewart Lockie. Australia: Pluto Press.

Logan, Mary E. and Helen Huntley. 2002. "Gender and Power in the Research Process." *Women's Studies International Forum* 24 (6):623-635.

MacDonald, Martha. 2005. "Lessons and Linkages: Building a Framework for Analysing the Relationships Between Gender, Globalization and the Fisheries." In *Changing Tides: Gender, Fisheries and Globalization*, edited by Barbara Neis, Marian Binkley, Siri Gerrard and Maria C. Maneschy, 18-27. Halifax: Fernwood Publishing.

Marine Stewardship Council (MSC). 2006. *Annual Report 2005/06.* London: Marine Stewardship Council.

Matics, K., O. Poeu, and I. Siason. 2002. "Networks for Women and Gender in Fisheries Development in the Mekong Region and the Philippines." In *Global Symposium on Women in Fisheries: Sixth Asian Fisheries Forum*, edited by MJ Williams, NH Chao, PS Choo, K Matics, MC Nandeesha, M Shariff, I Siason, E Tech and JMC Wong, 55-65. Malaysia: ICLARM-The World Fish Centre.

Maurstad, Anita. 2000. "Trapped in Biology: An Interdisciplinary Attempt to Integrate Fish Harvesters' Knowledge Into Norwegian Fisheries Management." In *Finding Our Sea Legs: Linking Fishery People and their Knowledge with Science and Management*, edited by Barbara Neis and Lawrence Felt, 135-152. Newfoundland: ISER.

Maurstad, Anita, and Jan Sundet. 1998. "The Invisible Cod – Fishermen's and Scientists' Knowledge." In *Commons in a Cold Climate*, edited by Svein Jentoft, 167-184. Paris: UNESCO.

Maynard, Mary. 1994. "Methods, Practice and Epistemology: The Debate About Feminism and Research." In *Researching Women's Lives From a Feminist Perspective,* edited by Mary Maynard and June Purvis, 10-27. London: Taylor and Francis.

Maynard, Mary and June Purvis, editors. 1994. *Researching Women's Lives from a Feminist Perspective*. London: Taylor and Francis.

McAllister, Jim and Barbara Geno. 2004. "Australian Farm Inheritance-New Patterns of Legal Structure in Property Rights and Landholding." *Rural Society* 14 (2):178-194.

McCay, Bonnie. 2005. "Gender, Globalization and a Tragic Choice on Fogo Island, Newfoundland: The Human Rights Case." In *Changing Tides: Gender, Fisheries and Globalization*, edited by Barbara Neis, Marian Binkley, Siri Gerrard and Christina Manezy, 116-129. Halifax: Fernwood Publishing.

McGoodwin, James, Barbara Neis and Lawrence Felt. 2000. "Integrating Fishery People and their Knowledge into Fisheries Science and Resource Management: Issues, Prospects and Problems." In *Finding Our Sea Legs: Linking Fishery People and their Knowledge with Science and Management,* edited by Barbara Neis and Lawrence Felt, 249-264. Newfoundland: ISER.

McKay, E, M. Dimopoulos, and R. Baker. n.d. *Needs Assessment and Analysis of Women in the Seafood Industry*. Victoria: Department of Natural Resources and Environment.

McNeil, Maureen. 1993. "Dancing with Foucault: Feminism and Power-Knowledge." In *Up Against Foucault: Explorations of Some Tensions Between Foucault and Feminism*, edited by Caroline Ramazanoglu, 147-178. London: Routledge.

Medard, Modesta and Douglas C. Wilson. 1996. "Changing Economic Problems for Women in the Nile Perch Fishing Communities on Lake Victoria." *Anthropologica* 38:149-172.

Merriam, Sharan B. 1998. *Qualitative Research and Case Study Applications*. San Francisco: Jossey-Bass Publishers.

Meyer, Katherine and Linda M. Lobao. 1994. "Engendering the Farm Crisis: Women's Political Response in the USA." In *Gender and Rurality,* edited by Sarah Whatmore, Terry Marsden and Phillip Lowe, 69-87. London: David Fulton Publishers.

Mhone, Guy. 2005. "*Neoliberalism.*" Encyclopedia.com. Accessed March 10 2011. http://www.encyclopedia.com/topic/Neoliberalism.aspx

Mies, Maria. 1983. "Towards a Methodology for Feminist Research." In *Theories for Women's Studies*, edited by G Bowles and RD Klein, 117-139. London: Routledge and Kegan Paul.

—. 1991, "Women's Research Or Feminist Research? The Debate Surrounding Feminist Science and Methodology." In *Beyond Methodology,* edited by Mary M. Fonow and Judith A Cook, 60-84. Bloomington: Indiana University Press.

Mies, Maria and Vandana Shiva. 2004. "The Subsistence Perspective." In *The Feminist Standpoint Theory Reader: Intellectual and Political Controversies*, edited by Sandra Harding, 333-338. New York: Routledge.

Miles, Matthew B., and A. Michael Huberman. 1994. *Qualitative Data Analysis: An Expanded Sourcebook.* 2nd ed. California: SAGE.

Millington, P and WJ. Fletcher. 2008. *Geelong Revisited: from ESD to EBFM–Future Directions for Fisheries Management. Workshop Report.* FRDC 2008/057, Melbourne May 2008. Fisheries Occasional Publication No.52, November 2008. Western Australia: Department of Fisheries.

Moore, Henrietta. 1993. "The Difference within and the Difference Between." In *Gendered Anthropology*, edited by Teressa del Valle, 193-204. London: Routledge.

Moreton-Robinson, Aileen. 2000, *Talkin' Up to the White Woman*. Queensland: University of Queensland Press.

Morgan, David L. and Richard A. Krueger. 1993. "When to Use Focus Groups and Why." In *Successful Focus Groups: Advancing the State of the Art,* edited by David L. Morgan, 3-20. Newbury Park: SAGE.

Muir, Sonia. 2003. "Growing Women Through the "Grass Ceiling."" In *Setting the Agenda for Rural Women: Research Directions, Conference Proceedings and Recommendations*, edited by Rowan O'Hagan, Margaret Alston and Shelley Spriggs, 198-208. Wagga

Wagga: Centre for Rural Social Research.

Munk-Madsen, Eva. 1998a. "The Norwegian Fishing Quota System: Another Patriarchal Construction?" *Society and Natural Resources* 11:229-240.

—. 1998b. "From Common Property to All Male Property-Quota Distribution in Norwegian Fishing." In *Commons in a Cold Climate*, edited by Svein Jentoft, 145-163. Paris: UNESCO.

—. 2000. "Wife the Deckhand, Husband the Skipper." *Women's Studies International Forum* 23 (3):333-342.

Myers, Greg and Phil Macnaghten. 1999. "Can Focus Groups Be Analysed as Talk?" In *Developing Focus Group Research: Politics, Theory and Practice*, edited by Rosaline S. Barbour and Jenny Kitzinger, 173-186. London: SAGE.

MyriaD Consultants. 2000. *Fishing for Success: Women in the Seafood Community Final Summary Report*. Victoria: Rural Women's Network (DNRE).

—. n.d. *Workshop Report: 12th December 2000, Wonthaggi Workmen's Club. Fishing for Success: Women in the Seafood Community*.

Nadel-Klein, Jane. 1988. "A Fisher Laddie Needs a Fisher Lassie: Endogamy and Work in a Scottish Fishing Village." In *To Work and to Weep: Women in Fishing Economies*, edited by Jane Nadel-Klein and Donna Lee Davis, 190-210. Newfoundland: Memorial University of Newfoundland.

—. 2000. "Granny Baited the Lines: Perpetual Crisis and the Changing Role of Women in Scottish Fishing Communities." *Women's Studies International Forum* 23 (3):363-372.

Nadel-Klein, Jane and Donna Lee Davis. 1988. "Introduction, Gender in The Maritime Arena." In *To Work and to Weep: Women in Fishing Economies*, Edited By Jane Nadel-Klein and Donna Lee Davis, 1-17. Newfoundland: Memorial University of Newfoundland.

Nadel-Klein, Jane and Donna Lee Davis, editors. 1988, *To Work and to Weep: Women in Fishing Economies*. Newfoundland: Memorial University of Newfoundland.

Nandeesha, MC. and E. Tech. 2002. "Women in Fisheries Activities of the Asian Fisheries Society-Have they Been Able to Make An Impact?" In *Global Symposium on Women in Fisheries: Sixth Asian Fisheries Forum,* edited by MJ Williams, NH Chao, PS Choo, K Matics, MC Nandeesha, M Shariff, I Siason, E Tech and JMC Wong, 7-13. Malaysia: ICLARM-The World Fish Centre.

National Aquaculture Council (NAC). 2001. *Australian Aquaculture Yearbook 2001*. Melbourne: Executive Media.

—. 2002. *Australian Aquaculture Yearbook 2002*. Australia: National Australian Aquaculture Council.

National Competition Policy (NCP) 2002. *National Competition Policy Review of Commonwealth Fisheries Legislation.* Accessed June 10 2005. http://www.daff.gov.au/__data/assets/word_doc/0019/5761/ncp_report_sep02_final.doc

Natural Resources and Environment. n. d. *Women in Agriculture and Resource Management Register Toolkit*. Victoria: Natural Resources and Environment.

Neis, Barbara. 2000. "In the Eye of the Storm: Research, Activism and Teaching within the Newfoundland Fishery Crisis." *Women's Studies International Forum* 23 (3):287-298.

—. 2005. "Introduction." In *Changing Tides: Gender, Fisheries and Globalization*, edited by Barbara Neis, Marian Binkley, Siri Gerrard and Maria C. Maneschy, 1-13. Halifax: Fernwood Publishing.

Neis, Barbara and Lawrence Felt. 2000. "Introduction." In *Finding Our Sea Legs: Linking Fishery People and their Knowledge with Science and Management,* edited by Barbara Neis and Lawrence Felt, 11-23. Newfoundland: ISER.

Neis, Barbara and Maneschy, Maria C. 2005. "Moving Forward." In *Changing Tides: Gender, Fisheries and Globalization*, edited by Barbara Neis, Marian Binkley, Siri Gerrard and Maria C. Maneschy, 245-255. Halifax: Fernwood Publishing.

Neis, Barbara, Marian Binkley, Siri Gerrard and Maria C Maneschy, editors. 2005. *Changing Tides: Gender, Fisheries and Globalization.* Halifax: Fernwood Publishing.

Organisation for Economic Co-operation and Development (OECD). 1999. *DAC Guidelines for Gender Equality and Women's Empowerment In Development Co-operation.* Development Co-operation Guidelines Series. Accessed March 2 2006. http://www.oecd.org/dataoecd/56/46/28313843.pdf

O'Hagan, Rowan, Margaret Alston and Shelley Spriggs. 2003. *Setting the Agenda for Rural Women: Research Directions, Conference Proceedings* and Recommendations. Wagga Wagga: Centre for Rural Social Research.

O'Hara, Patricia. 1994. "Constructing the Future: Cooperation and Resistance Among Farm Women in Ireland." In *Gender and Rurality*, edited by Sarah Whatmore, Terry Marsden and Phillip Lowe, 50-68. London: David Fulton Publishers.

Okraku, Ishmael O. 1975. Fishing and Fertility: A Study of a Nova Scotia

Fishing Village." *Social Biology* 22 (4):326-337.

Palsson, Gisli. 2000. "'Findings One's Sea Legs': Learning the Process of Enskilment, and Integrating Fishers' and their Knowledge into Fisheries Science and Management." In *Finding Our Sea Legs: Linking Fishery People and their Knowledge with Science and Management,* edited by Barbara Neis and Lawrence Felt, 26-40. Newfoundland: ISER.

Phoca, Sophia and Rebecca Wright. 1999. *Introducing Postfeminism.* Sydney: Allen and Unwin.

Pini, Barbara. 1998. "The Emerging Economic Rationalist Discourse on Women and Leadership in Australian Agriculture." *Rural Society* 8 (3):223-235.

—. 2002a. "The Exclusion of Women From Agri-Political Leadership: A Case Study of the Australian Sugar Industry." *Sociologia Ruralis* 42 (1):65-76.

—. 2002b. "Focus Groups, Feminist Research and Farm Women: Opportunities for Empowerment in Rural Social Research." *Journal of Rural Studies* 18:339-351.

—. 2003a. "Putting the Theory into Practice: A Description of a Feminist Research Project in Rural Sociology." In *Setting the Agenda for Rural Women: Research Directions and Conference Proceedings*, edited by Rowan O'Hagan, Margaret Alston and Shelley Spriggs, 110-125. Wagga Wagga: Centre for Rural Social Research.

—. 2003b. "Sheep, Shadows and Silly Saints: Constructing Women in Agricultural Leadership." *Rural Society* 13 (2):193-211.

—. 2004. "Managerial Masculinites in the Australian Sugar Industry." *Rural Society* 14 (1):22-35.

—. 2005a. "Farm Women: Driving Tractors and Negotiating Gender." International *Journal of Sociology of Agriculture and Food* 13 (1):1-12.

—. 2005b. "The Third Sex: Women Leaders in Australian Agriculture." *Gender, Work and Organisation* 12 (1):73-88.

Plumwood, Val. 1993. *Feminism and the Mastery of Nature*. London: Routledge.

Poiner, Gretchen. 1990. *The Good Old Rule: Gender and Other Power Relationships in a Rural Community*. Australia: Sydney University Press.

—. 1994. "Landscape with Figures: Femininity and Rural Views." In *Country Women At the Crossroads: Perspectives on the Lives of Rural Australian Women in the 1990s*, edited by Margaret-Ann Franklin, Leonie M. Short and Elizabeth K. Teather, 51-63. Armidale:

University of New England Press.

Poole, Marilyn. 1997. "The Search for Liberation: Feminism." In *Sociology: Australian Connections*, edited by Ray Jureidini, Sue Kenny and Marilyn Poole, 94-116. Australia: Allen and Unwin.

Porter, Marilyn. 1983. ""Women and Old Boats": The Sexual Division of Labour in a Newfoundland Outport." In *The Public and the Private*, edited by Eve Gamarnikow, David H. Morgan, June Purvis and Daphne Taylorson, 91-106. London: Heinemann.

—. 1988. "Mothers and Daughters: Linking Women's Life Histories in Grand Bank, Newfoundland, Canada." *Women's Studies International Forum* 11 (6):545-558.

—. 1991. "Time, the Life Course and Work In Women's Lives: Reflections from Newfoundland." *Women's Studies International Forum* 14 (1-2):1-13.

Porter, Terry B. n.d. "Identity Subtexts in the Discursive Construction of Sustainability." Accessed December 2 2006. http://www.mngt.waikato.ac.nz/ejrot/Vol9_1/Porter.pdf.

Power, Nicole G. 2000. "Women Processing Workers as Knowledgeable Resource Users: Connecting Gender, Local Knowledge and Development In the Newfoundland Fishery." In *Finding Our Sea Legs: Linking Fishery People and their Knowledge with Science and Management*, edited by Barbara Neis and Lawrence Felt, 189-203. Newfoundland: ISER.

Power, Nicole G. and Donna Harrison. 2005. "The Power of Gender Ideology In the Face of Resource Decline In Newfoundland, Canada." In *Changing Tides: Gender, Fisheries and Globalization*, edited by Barbara Neis, Marian Binkley, Siri Gerrard and Christina Manezy, 229-241. Halifax: Fernwood Publishing.

Power, Nicole G. 2005. "The "Modern Fisherman": Masculinity In Crisis Or Resilient Masculinity?" *Canadian Woman Studies* 24 (4):102-107.

Primary Industries Standing Committee (PISC). 2003. *A Vision for Change: National Plan for Women in Agriculture and Natural Resource Management, Third Annual Implementation Report 2000-2001*. Canberra: AFFA.

Quible, Zane K. 1998. "A Focus on Focus Groups." *Business Communication Quarterly* 61 (2):28-38.

Ram, Kalpana. 1991. *Mukkuvar Women: Gender, Hegemony and Capitalist Transformation in a South Indian Fishing Community.* Australia: Allen and Unwin.

Ramazanoglu, Caroline. 1993. "Introduction." In *Up Against Foucault: Explorations of Some Tensions Between Foucault and Feminism*, edited by Caroline Ramazanoglu, 1-28. London: Routledge.

Ramazanoglu, Caroline, and Janet Holland. 1993. "Women's Sexuality and Men's Appropriation of Desire." In *Up Against Foucault: Explorations of Some Tensions Between Foucault and Feminism*, edited by Caroline Ramazanoglu, 239-264. London: Routledge.

Rana, K. and PS. Choo. 2002. "Women in Fisheries in the European Union." In *Global Symposium on Women in Fisheries: Sixth Asian Fisheries Forum,* edited by MJ Williams, NH Chao, PS Choo, K Matics, MC. Nandeesha, M Shariff, I Siason, E Tech and JMC Wong, 191-197. Malaysia: ICLARM-The World Fish Centre.

Ransom, Janet. 1993. "Feminism, Difference and Discourse: The Limits of Discursive Analysis for Feminism." In *Up Against Foucault: Explorations of Some Tensions Between Foucault and Feminism*, edited by Caroline Ramazanoglu, 123-146. London: Routledge.

Reinharz, Shulamit. 1992, *Feminist Methods in Social Research.* New York: Oxford University Press.

—. 1997. "Who Am I? The Need for a Variety of Selves in the Field." In *Reflexivity and Voice*, edited by Rosanna Hertz, 3-20. California: SAGE.

Rickson, Sarah T. 1997. "Outstanding in their Field: Women in Agriculture." *Current Sociology* 45 (2):91-133.

RIRDC/ DPIE. 1998a. *Missed Opportunities-Harnessing the Potential of Women in Australian Agriculture, Volume 1. Social survey and analysis, (Social survey and analysis: Volume 1).* Canberra: RIRDC/ DPIE.

—. 1998b. *Missed Opportunities-Harnessing the Potential of Women in Australian Agriculture, Volume 2.Economic module-Estimating the current and potential contribution of women in agriculture.* Canberra: RIRDC/ DPIE.

Robins, C and M. Sachse. 1994. "Time and Tides: A History of the Fishery." In *Australia's Northern Prawn Fishery*, edited by Peter C. Pownall, 1-22. Australia: NPF25.

Roepstorff, Andreas. 2000. "The Double Interface of Environmental Knowledge: Fishing for Greenland Halibut." In *Finding Our Sea Legs: Linking Fisheries People and their Knowledge with Science and Management*, edited by Barbara Neis and Lawrence Felt, 165-188. Newfoundland: ISER.

Rose, Howard. 1991. "Case Studies." In *Handbook for Research Students in the Social Sciences*, edited by Graham Allan and Chris Skinner, 190-202. London: The Falmer Press.

Rowe, Sherrylynn, and George Feltham. 2000. "East Port Peninsula

Lobster Conservation: Integrating Harvesters' Local Knowledge and Fisheries Science for Resource Co-Management." In *Finding Our Sea Legs: Linking Fishery People and their Knowledge with Science and Management*, edited by Barbara Neis and Lawrence Felt, 236-246. Newfoundland: ISER.

Rubinoff, Janet A. 1999. "Fishing for Status: Impact of Development on Goa's Fisherwomen." *Women's Studies International Forum* 22 (6):630-644.

Ruddick, Sara. 2004. "Maternal Thinking as a Feminist Standpoint." In *The Feminist Standpoint Theory Reader: Intellectual and Political Controversies*, edited by Sandra Harding, 161-168. New York: Routledge.

Sachs, Carolyn. 1983. *The Invisible Farmers: Women in Agricultural Production.* New Jersey: Rowan and Allanheld.

—. 1994. "Rural Women's Environmental Activism in the USA." In *Gender and Rurality*, edited by Sarah Whatmore, Terry Marsden and Phillip Lowe, 117-136. London: David Fulton Publishers.

—. 1996. *Gendered Fields: Rural Women, Agriculture and Environment.* USA: Westview Press.

Schirmer, J. 2005. *A Social Assessment Handbook for Use By Australian Fisheries Managers in ESD Assessment and Monitoring, ESD Reporting and Assessment Subprogram Publication No. 14.* Canberra, ACT: Fisheries Research and Development Corporation and Bureau of Rural Sciences.

Schirmer, J. and A. Casey. 2005. *Social Assessment Handbook: A Guide to Methods and Approaches for Assessing the Social Sustainability of Fisheries in Australia.* Canberra, ACT: Bureau of Rural Sciences.

Schirmer, J. and J Pickworth. 2005a. *Social Impacts of the South Australian Marine Scalefish Fishery, FRDC ESD Reporting and Assessment Subprogram Publication No. 10.* Canberra, ACT: Fisheries Research and Development Corporation and Bureau of Rural Sciences.

Schirmer, J and J Pickworth. 2005b. *Social Assessment of Commercial Fishing in the East Gippsland region, FRDC ESD Reporting and Assessment Subprogram Publication No. 11.* Canberra, ACT: Fisheries Research and Development Corporation and Bureau of Rural Sciences.

Shariff, M. 2002. "Asian Fisheries Society: An Historical Perspective of Women in Fisheries." In *Global Symposium on Women in Fisheries: Sixth Asian Fisheries Forum,* edited by MJ Williams, NH Chao, PS Choo, K Matics, MC Nandeesha, M Shariff, I Siason, E Tech and JMC Wong, 3-15. Malaysia: ICLARM-The World Fish Centre.

Shelly, AB. and M. D'Costa. 2002. "Women in Aquaculture: Initiatives of

Caritas Bangladesh." In *Global Symposium on Women in Fisheries: Sixth Asian Fisheries Forum*, edited by MJ Williams, NH Chao, PS Choo, K Matics, MC Nandeesha, M Shariff, I Siason, E Tech and JMC Wong. Malaysia: ICLARM-The World Fish Centre.

Sheridan, Alison. 1994. "Women in Agriculture-Where Are They?" In *Country Women At the Crossroads: Perspectives on the Lives of Rural Australian Women in the 1990s*, edited by Margaret A. Franklin, Leonie M. Short and Elizabeth K. Teather, 18-24. Armidale: University of New England Press.

Sheridan, Alison and Fiona Haslam McKenzie. 2009. *Revisiting Missed Opportunities-Growing Women's Contribution to Agriculture.* Canberra, ACT: RIRDC.

Shortall, Sally. 1994. "Farm Women's Groups: Feminist Or Farming Or Community Groups, Or New Social Movements?" *Sociology* 28 (1):279-291.

—. 1999. *Women and Farming: Property and Power.* Great Britain: Macmillan Press.

—. 2000. "In and Out of the Milking Parlour: A Cross-National Comparison of Gender, the Dairy Industry and the State." *Women's Studies International Forum* 23 (2):247-257.

—. 2001. "Women in the Field: Women, Farming and Organizations." *Gender, Work and Organization* 8 (2):164-181.

—. 2003. "An International Perspective on Rural Women's Research: Identifying Common Themes." In *Setting the Agenda for Rural Women: Research Directions, Conference proceedings and recommendations,* edited by Rowan O'Hagan, Margaret Alston and Shelley Spriggs, 52-64. Wagga Wagga: Centre for Rural Social Research.

—. 2004a. "The Broad and Narrow: Case Studies and International Perspectives on Farm Women's Research." *Rural Society* 14 (2):112-116.

—. 2004b. "Social Or Economic Goals, Civic Inclusion Or Exclusion? An Analysis of Rural Development, Theory and Practice." *Sociologia Ruralis* 44 (1):109-123.

—. 2005. "Political Climate and Gender Relations: Farm Women, Property Rights and Equality in Northern Ireland." In *Critical Studies in Rural Gender Issues*, edited by Jo Little and Carolyn Morris, 90-103. England: Ashgate.

Siason, I., E. Tech, KI. Matics, PS. Choo, M. Shariff, ES. Heruwati, T. Susilowati, N. Miki, AB. Shelly, KG. Rajabharsi, R. Ranjit, PPGN. Siriwardena, MC. Nandeesha and M. Sunderarajan. 2002. "Women in

Fisheries in Asia." In *Global Symposium on Women in Fisheries: Sixth Asian Fisheries Forum,* edited by MJ Williams, NH Chao, PS. Choo, K Matics, MC Nandeesha, M Shariff, I Siason, E Tech and JMC Wong, 21-49. Malaysia: ICLARM-The World Fish Centre.

Skaptadottir, Unnur D., and Hulda Proppe. 2005. "Global Processes, Localities and Gender Identities: A Feminist Perspective on Changes in Icelandic Fisheries." In *Changing Tides: Gender, Fisheries and Globalization*, edited by Barbara Neis, Marian Binkley, Siri Gerrard and Christina Manezy, 152-165. Halifax: Fernwood Publishing.

Skaptadottir, Unnur D. 1996. "Gender Construction and Diversity in Icelandic Fishing Communities." *Anthropologica* 38:271-287.

—. 2000. "Women Coping with Change in An Icelandic Fishing Community: A Case Study." *Women's Studies International Forum* 23 (3):311-321.

Skeggs, Beverly. 1997. *Formations of Class and Gender.* London: SAGE.

Smith, Dorothy. 1987. "Women's Perspective as a Radical Critique of Sociology." In *Feminism and Methodology*, edited by Sandra Harding, 84-97. USA: Indiana University and Open University Press.

—. 1996. "Contradictions for Feminist Social Scientists." In *Feminism and Social Change: Bridging Theory and Practice*, edited by Heidi Gottfried, 46-69. Urbana: University of Illinois Press.

—. 2004a. "Women's Perspective as a Radical Critique of Sociology." In *The Feminist Standpoint Theory Reader: Intellectual and Political Controversies*, edited by Sandra Harding, 21-34. New York: Routledge.

—. 2004b. "Comment on Heckman's "Truth and Method: Feminist Standpoint Theory Revisited." In *The Feminist Standpoint Theory Reader: Intellectual and Political Controversies*, edited by Sandra Harding, 263-268. New York: Routledge.

Smith, EE. 1988. "The Right to Choice: Power and Decision Making." In *To Work and to Weep: Women in Fishing Economies*, edited by Jane Nadel-Klein and Dona Lee Davis, 279-292. Newfoundland: Memorial University of Newfoundland.

Smith, H. and D. Sullivan. 2001. *Seafood Women Embrace: Best Practice Decision Making Report.* Australia: WINSC.

Soper, Kate. 1993. "Productive Contradictions." In *Up Against Foucault: Explorations of Some Tensions Between Foucault and Feminism*, edited by Caroline Ramazanoglu, 29-50. London: Routledge.

Spalter-Roth, Roberta and Heidi Hartmann. 1999. "Small Happiness: The Feminist Struggle to Integrate Social Research with Social Activism." In *Feminist Approaches to Theory and Method*, edited by Sharlene

Hesse-Biber, Christina Gilmartin and Robin Lydenberg, 333-348. New York and Oxford: Oxford University Press.

Stacey, Judith. 1988. "Can There Be a Feminist Ethnography?" *Women's Studies International Forum* 11 (1):21-27.

Stake, Robert E. 1995. The *Art of Case Study Research*. Thousand Oaks: SAGE.

—. 2000. "The Case Study Method in Social Inquiry." In *Case Study Method*, edited by Roger Gomm, Martyn Hammersely and Peter Foster, 19-26. London: SAGE.

Standing Committee on Agriculture and Resource Management (SCARM). 1998. *A Vision for Change: National Plan for Women in Agriculture and Resource Managem*ent. Canberra, ACT: SCARM.

—. 2001. *A Vision for Change: National Plan for Women in Agriculture and Resource Management, Second Implementation Report 1999-2000*. Canberra, ACT: SCARM.

Stanley, Liz, and Sue Wise. 1993. *Breaking Out Again: Feminist Ontology and Epistemology*. 2nd ed. London: Routledge and Kegan Paul.

Stella, Leonie. 1996. "No Place for Wimps: Working on Western Australian Trawlers." *Anthropologica* 38:173-193.

—. 1998. *Trawling Deeper Seas: The Gendered Production of Seafood in Western Australia*. Perth, Western Australia: Murdoch University. Unpublished PhD thesis.

Symes, David. 1996. "Introduction." In *Fisheries Management in Crisis*, edited by Kevin Crean and David Symes, 3-18. Oxford: Fishing News Books.

Szala-Meneok, Karen. 1996. "Craft Development and Development Through Crafts: Adaptive Strategies of Labrador Women in a Changing Fishery." *Anthropologica* 38:249-270.

Teather, Elizabeth K. 1992. "Remote Rural Women's Ideologies, Spaces and Networks: Country Women's Association of New South Wales 1922-1992." *ANZJS* 28 (3):369-390.

—. 1994a. "Contesting Rurality: Country Women's Social and Political Networks." In *Gender and Rurality*, edited by Sarah Whatmore, Terry Marsden and Phillip Lowe, 31-50. London: David Fulton Publishers.

—. 1994b. "CWA At the Crossroads." In *Country Women At the Crossroads: Perspectives on the Lives of Rural Australian Women in the 1990s*, edited by Margaret-Ann Franklin, Leonie M. Short and Elizabeth K. Teather, 134-140. Armidale: University of New England Press.

—. 1996a. "Rural Women's Self-Concepts and Aspirations as Members of Selected Voluntary Organisations in New Zealand, Australia and

Canada." *New Zealand Geographer* 52 (2):35-45.

—. 1996b. "Farm Women in Canada, New Zealand and Australia Redefine their Rurality." *Journal of Rural Studies* 12 (1):1-14.

—. 1998. "The Double Bind: Being Female and Being Rural: A Comparative Study of Australia, New Zealand and Canada." *Rural Society* 8 (3):209-223.

Thiessen, Victor, Anthony Davis and Svein Jentoft. 1992. "The Veiled Crew: An Exploratory Study of Wives' Reported and Desired Contributions to Coastal Fisheries Enterprises in Northern Norway and Nova Scotia." *Human Organization* 51 (4):342-352.

Thompson, Paul. 1985. "Women in Fishing: The Roots of Power Between the Sexes." Comparative Studies in Society and History 27:3-32.

Thompson, Paul, Tony Wailey and Trevor Lummis. 1983. *Living the Fishing*. London: Routledge and Kegan Paul.

Tonts, Matthew, and Alan Black. 2002. *The Impacts of Changing Farm Business Structures on Rural Communities, Publication No. 02/027 No. ECU-10A*. Australia: Rural Industries Research and Development Corporation.

Trauger, Amy. 2004. "Because they Can Do the Work: Women Farmers in Sustainable Agriculture in Pennsylvania, USA." *Gender, Place and Culture* 11 (2):289-307.

Tsamenyi, B. Martin and Kwame Mfodwo. 2001. *Towards Greater Indigenous Participation in Australian Commercial Fisheries: Some Policy Issues*. Australia: ATSIC.

UN Earthwatch. 2006. "Crisis in Ocean Fisheries." *UNEP/DEWA/ Earthwatch*. Accessed December 2 2006.
earthwatch.unep.net/emergingissues/oceans/oceanfisheries.php-32k

United Nations. 1995. Fourth World Conference on Women: Action for Equality, Development and Peace, Beijing, China September 1995. Accessed March 20 2006.
www.un.org/womenwatch/daw/beijing/index.html-14k

United Nations Population Fund (UNFPA). n.d. Women in Power and Decision-Making. Accessed February 6 2006.
www.unfpa.org/intercenter/beijing/power.htm

Voorhaar, R. 2000. "Network to Unite Fishing's On-Shore Partners." *The Age*, 22 September.

Voyce, Malcolm. 1994. "Testamentary Freedom: Patriarchy and inheritance of the Family Farm." *Sociologia Ruralis* 34 (1):71-83.

—. 1997. "Inheritance and Family Farm Ownership." In *A Legacy Under Threat? Family Farming in Australia*, edited by J Lees, 199-213. Australia: University of New England Press.

Waitt, Gordon and Kate Hartig. 2000. "Ecologically Sustainable Fishing in Theory and Practice: Individual Transferable Quotas in Australia's South East Trawl Fishery." *Australian Geographer* 31 (1):87-114.

Waitt, Gordon and Kate Hartig. 2005. "All At Sea. Rethinking Fisher's Identity in Australia." *Gender, Place and Culture* 12 (4):403-418.

Waldby, Catherine. 1995. "Feminism and Method." In *Transitions: New Australian Feminisms*, edited by Barbara Caine and Rosemary Pringle, 15-29. Australia: Allen and Unwin.

Walker, Barbara LE. 2001. "Sisterhood and Seine-Nets: Engendering Development and Conservation in Ghana's Marine Fishery." *Professional Geographer* 53 (2):160-177.

—. 2002. "Engendering Ghana's Seascape: Fanti Fishtraders and Marine Property in Colonial History." *Society and Natural Resources* 15:389-407.

Ward, Claudine. 2004. *""Do You Go Out on the Boat Too, Luv?" Stories From Women in the Gulf of Carpentaria Fishery."* Karumba: Claudine Ward.

Waring, Marilyn. 1998. "A Question of Value." In *Australian Rural Women Towards 2000,* edited by Margaret Alston, 5-15. Wagga Wagga: Centre for Rural Social Research.

Weedon, Chris. 1989. *Feminist Practice and Poststructuralist Theory.* Oxford: Blackwell.

—. 1999. *Feminism, Theory and the Politics of Difference.* Oxford: Blackwell.

Weeks, Kathi. 2004. "Labor, Standpoints and Feminist Subjects." In *The Feminist Standpoint Theory Reader: Intellectual and Political Controversies*, edited by Sandra Harding, 181-194. New York: Routledge.

Weeratunge, Nireka and Katherine Snyder. 2009. *Gleaner, Fisher, Trader, Processor: Understanding Gendered Employment in The Fisheries and Aquaculture Sector*, Paper presented at the FAO-IFAD-ILO Workshop on Gaps, trends and current research in gender dimensions of agricultural and rural employment: differentiated pathways out of poverty, Rome, 31 March - 2 April 2009. Malaysia/Malawi: World Fish Centre.

Weiner, Gaby. 1993. "The Gendered Curriculum-Producing the Text: Developing a Poststructural Feminine Analysis." Paper presented at the annual Conference of the Australian Association for Research in Education, Perth, December 1993. Accessed March 18 2006. www.aare.edu.au/93pap/weing93333.txt-62k.

Whatmore, Sarah. 1991. *Farming Women: Gender, Work and Family*

Enterprise. London: Macmillan Academic and Professional Ltd.

Whatmore, Sarah, Terry Marsden and Phillip Lowe, editors. 1994. *Gender and Rurality.* London: David Fulton Publishers.

Whiting, Emma and Henry Hough. 1953. *Whaling Wives.* Boston: Houghton Mifflin.

Wilkinson, Jane. 1998a. "Prejudices, Preconceptions Or Progress in Agriculture? Rural Non English Speaking Background Women and Leadership." *Rural Society* 8 (3):235-257.

—. 2004. *NSW Fishing Industry: Changes and Challenges in the Twenty-First Century, Briefing Paper No 11/04.* NSW: NSW Parliamentary Library Research Service.

Wilkinson, Jane and Margaret Alston. 1999. *Let's Walk the Talk Inside the Department.* Wagga Wagga: Centre for Rural Social Research.

Wilkinson, Sue. 1998b. "Focus Groups in Feminist Research: Power, Interaction, and the Co-Construction of Meaning." *Women's Studies International Forum* 21 (1):111-125.

—. 1999. "How Useful Are Focus Groups in Feminist Research?" In *Developing Focus Group Research: Politics, Theory and Practice,* edited by Rosaline S. Barbour and Jenny Kitzinger, 64-79. London: SAGE.

Williams, Meryl J., SB. Williams and Poh Sze Choo. 2002. "From Women in Fisheries to Gender and Fisheries." In *Global Symposium on Women in Fisheries: Sixth Asian Fisheries Forum,* edited by MJ Williams, NH Chao, PS Choo, K Matics, MC Nandeesha, M Shariff, I Siason, E Tech and JMC Wong, 145-155. Malaysia: ICLARM-The World Fish Centre.

Williams, Julie. 1992. *The Invisible Farmer: A Report on Australian Farm Women.* Canberra: DPIE.

Williams, Meryl J. 2002. "Women in Fisheries: Pointers for Development." In *Global Symposium on Women in Fisheries Sixth Asian Fisheries Forum,* edited by Meryl J Williams, Nai-Hsien Chao, Poh Sze Choo, K Matics, MC Nandeesha, M Shariff, I Siason, E Tech and JMC Wong, vii-1. Malaysia: ICLARM-The World Fish Centre.

Williams, Meryl J., NH. Chao, Poh Sze Choo, K. Matics, MC. Nandeesha, M. Shariff, I. Siason, E. Tech and JMC Wong, editors. 2002. *Global Symposium on Women in Fisheries Sixth Asian Fisheries Forum.* Malaysia: ICLARM-The World Fish Centre.

Williams, Meryl J., VP. Nandeesha, E. Corral and Poh Sze Choo. 2001. *International Symposium on Women in Asian Fisheries: Fifth Asian Fisheries Forum.* Malaysia: ICLARM.

Williams, SB. 2002. "Making Each and Every African Fisher Count: Women Do Fish." In *Global Symposium on Women in Fisheries: Sixth*

Asian Fisheries Forum, edited by MJ Williams, NH Chao, PS Choo, K Matics, MC Nandeesha, M Shariff, I Siason, E Tech and JMC Wong. Malaysia: ICLARM-The World Fish Centre.

Willot, Sara and Christine Griffin. 1997. "Wham Bam, am I a Man?: Unemployed Men Talk about Masculinities." *Feminism and Psychology* 7(1):107-128.

Women in Agriculture International Conference Committee. 1995. *Women in Agriculture: Farming for Our Future, 1994 International Post-Conference Proceedings.* Women in Agriculture International Conference Committee.

Women's Industry Network Seafood Community (WINSC). 2001. *Annual Report 2000-2001*. South Australia: WINSC.

—. n.d. *Corporate Objectives and Strategic Priorities 2000-2003*. South Australia: WINSC.

World Commission on Environment and Development (WCED). 1990. *Our Common Future*, Australian ed. Melbourne: Oxford University Press.

Yater, Luz R. 1982. "The Fisherman's Family: Economic Roles of Women and Children." In *Small-Scale Fisheries of San Miguel Bay, Philippines: Social Aspects of Production and Marketing*, edited by Connor Bailey, 42-51. Philippines: Institute of Fisheries Development and Research, University of the Philippines in the Visayas; International Centre for Living Aquatic Resources Management; United Nations University.

Yin, Robert K. 2003a. *Case Study Research: Design and Methods*. 3rd ed. USA: Sage Publications.

—. 2003b. *Applications of Case Study Research*, 2nd ed. USA: Sage Publications.

Yodanis, Carrie L. 2000. Constructing Gender and Occupational Segregation: A Study of Women and Work in Fishing Communities." *Qualitative Sociology* 23 (3):267-290.

Zwanenburg, Kees, Patricia King and Paul Fanning. 2000. "Fishermen and Scientists Research Society: A Model for Incorporating Fishers and their Knowledge into Stock Assessment." In *Finding Our Sea Legs: Linking Fisheries People and their Knowledge with Science and Management,* edited by Barbara Neis and Lawrence Felt, 124-132. Newfoundland: ISER.

Zweifel, Helen. 1997. "The Gendered Nature of Biodiversity Conservation." *NWSA Journal* 9 (3):107-123.

INDEX